GUIDELINES FOR

Improving Plant Reliability through Data Collection and Analysis

Publications Available from the
CENTER FOR CHEMICAL PROCESS SAFETY
of the
AMERICAN INSTITUTE OF CHEMICAL ENGINEERS

CCPS Guidelines Series

Guidelines for Improving Plant Reliability through Data Collection and Analysis
Guidelines for Pressure Relief and Effluent Handling Systems
Guidelines for Design Solutions for Process Equipment Failures
Guidelines for Safe Warehousing of Chemicals
Guidelines for Postrelease Mitigation in the Chemical Process Industry
Guidelines for Integrating Process Safety Management, Environment, Safety, Health, and Quality
Guidelines for Use of Vapor Cloud Dispersion Models, Second Edition
Guidelines for Evaluating Process Plant Buildings for External Explosions and Fires
Guidelines for Writing Effective Operations and Maintenance Procedures
Guidelines for Chemical Transportation Risk Analysis
Guidelines for Safe Storage and Handling of Reactive Materials
Guidelines for Technical Planning for On-Site Emergencies
Guidelines for Process Safety Documentation
Guidelines for Safe Process Operations and Maintenance
Guidelines for Process Safety Fundamentals in General Plant Operations
Guidelines for Chemical Reactivity Evaluation and Application to Process Design
Tools for Making Acute Risk Decisions with Chemical Process Safety Applications
Guidelines for Preventing Human Error in Process Safety
Guidelines for Evaluating the Characteristics of Vapor Cloud Explosions, Flash Fires, and BLEVEs
Guidelines for Implementing Process Safety Management Systems
Guidelines for Safe Automation of Chemical Processes
Guidelines for Engineering Design for Process Safety
Guidelines for Auditing Process Safety Management Systems
Guidelines for Investigating Chemical Process Incidents
Guidelines for Hazard Evaluation Procedures, Second Edition with Worked Examples
Plant Guidelines for Technical Management of Chemical Process Safety, Revised Edition

Continued at the end of this book

GUIDELINES FOR

Improving Plant Reliability through Data Collection and Analysis

CENTER FOR CHEMICAL PROCESS SAFETY
of the
AMERICAN INSTITUTE OF CHEMICAL ENGINEERS

3 Park Avenue, New York, New York 10016-5901

Library of Congress Cataloging-in Publication Data
Guidelines for improving plant reliability through data collection and analysis / Center for Chemical Process Safety of the American Institute of Chemical Engineers
p. cm.
Includes bibliographical references and index.
ISBN 0-8169-0751-X
1. Chemical process control—Statistical methods. 2. Reliability (Engineering)—Statistical methods. I. American Institute of Chemical Engineers. Center for Chemical Process Safety.
TP155.75.G85 1998 98-39578
660´.2815—dc21 CIP

Contents

Preface

The unique value of this book is that it is the defining document for an industry-wide Plant and Equipment Reliability Database. The intent, structure, and implementation of the database are described fully within the book and its appendices. This Guideline book, developed for the Center for Chemical Process Safety (CCPS), is designed as a text to be used by operating and maintenance staff, reliability engineers, design engineers, and risk analysts. It treats the broad topic of equipment reliability, but also provides details about the design and operation of a process-plant/equipment reliability database.

The major objective of this book is to lay the foundation for an industry-wide reliability database. In fulfilling this objective, the book satisfies three auxiliary objectives. The first is to document and explain the theory of reliability data, including failure rates and the data structure employed by CCPS to accomplish its goals. The second is to demonstrate the usefulness of quality data by presenting worked examples. The text emphasizes that data needs are driven by analyses that provide added value. The book will help to illustrate this point. It will also help the reader understand how to actually carry out the analyses. The third objective is to provide an overview for the necessary quality assurance that must be implemented by both the plants maintaining a local database and a centralized database administrator to whom data is submitted for aggregation and distribution.

After an introduction to the purpose and basic operating concepts of the database in Chapter 1, the necessary background and terminology for reliability data analysis are developed in Chapters 2 and 3. Chapter 4 provides example applications of developing reliability data. Chapter 5 covers the details behind the structure of the CCPS database. Chapter 6 deals with quality assurance for the database.

A set of appendices provides specific information about database operating guidelines, pick lists, and a procedure for developing future equipment taxonomy information.

Acknowledgements

The Center for Chemical Process Safety (CCPS) wishes to thank all the company representatives of the sponsor companies of the Process Equipment Reliability Database Project who provided the invaluable guidance in the development of this protocol. In particular, the leadership of Mr. Hal Thomas of Air Products and Chemicals as Project Chairman, was instrumental in the completion of this project. CCPS also wishes to express its appreciation to the members of the equipment subcommittees for the development of the database taxonomies.

The company representatives include the following individuals:

Hal Thomas (Air Products & Chemicals)
Ed Zamejc (Amoco Corporation),
Donnie Carter and Patrick McKern (ARCO)
Malcolm Hide (BOC Group)
Colin Wills and Bobby Slinkard (BP Oil International)
Richard King and Peter Hughes (Caltex Services Corp)
Ashok Rakhe and Dan Long (Celanese Limited)
Wilbert Lee (Chevron Research and Technology Co.)
Ted Bennett (Dow Chemical Co)
Roy Schuyler (DuPont Engineering)
Tom Whittemore and Mark Templeton (Eastman Chemical Co.)
Robert Motylenski (Exxon Research & Engineering Co)
Kumar Bhimavarapu (Factory Mutual Research Corp)
Jack Dague (GE Plastics)
Larry Meier and Keith Burres (Hartford Steam Boiler Inspection and Insurance Co)
Gary Helman (Hercules, Inc)
Malcolm Preston, John Nigel, and Keith Buchwald (ICI)
Adrian Balda (Intevep S.A.)

Tetsuo Shiro and Morio Fujimura (Mitsubishi Chemical Co.)
Carlton Jensen (Phillips Petroleum Co)
William Angstadt and Richard Pettigrew (Rohm and Haas Co)
Harley Tripp, John Reynolds, and Mike Drossjack (Shell Oil Co)
Rob Mann and Darrel Alger (Syncrude Canada, Ltd)
Tommy Martin (Texaco USA)
Douglas Cramer (Westinghouse Savannah River Co).

The Equipment Subcommittee Chairman included:

Paul Altpeter—Compressors (Air Products and Chemicals Co)
William Angstadt—Instrumentation (Rohm and Haas Co)
Ted Bennett—Furnaces (Dow Chemical Co)
Douglas Ferguson—Vessels (DuPont Engineering)
Tommy Martin—Piping (Texaco, USA)
Larry Meier—Heat Exchangers (Hartford Steam Boiler Inspection and Insurance Co.)
Tom Whittemore, Jr.—Pumps (Eastman Chemical Co)
Ed Zamejc—Pressure Reliving Devices (Amoco Corporation).

Det Norske Veritas (USA) Inc. (DNV), Houston, Texas, was the contractor who prepared these Guidelines and is providing the software development. Bernie Weber was DNV's Project Manager. The principal authors were Mike Moosemiller, and Bernie Weber. A special thanks to Anne McKinney for her technical assistance and to Kathy Dubiel and Kelly Holmstrom who provided dedication and skill in manuscript preparation.

This project is the fruition of equipment reliability database projects starting in 1987 championed by Les Wittenberg, CCPS and managed by Joseph Metallia as a Staff Consultant for CCPS. Les and Joe struggled with how to get a database going over much criticism. In late 1997 Joe Metallia turned over the project management to a new staff consultant, David Belonger. Andrew Wolford of DNV provided many creative ideas and lived through the lean years. His contributions to obtain sponsors were key to the on-going database.

We gratefully acknowledge the comments and suggestions submitted by the following companies and peer reviewers: Lucia Dunbar (AEA Technology), G. Bradley Chadwell (Battelle Memorial Institute), Daniel Crowl (Michigan Technological University), Robert L. Post (Rohm and Haas Company), Brian D. Kelly (Syncrude Canada Ltd).

1 Introduction

1.1. Background

Quality failure rate data have long been a desire within the Chemical Process Industry. Unfortunately, too often the emphasis has been on the collection of data (which in and of itself is useless) rather than on value-added uses of data. This implies that one must answer the question of what to do with the data once it is collected. To some extent this is done, but generally the effort has been cut short and has mainly addressed perceived data requirements, regulatory requirements, and data that make day-to-day work assignments more efficient, or provides proof of work performed. The type of information that would allow more effective continuous improvement is too often included without any real thought other than, "sounds like we should have it," or "we might need it someday."

The major objective of this book is to lay the foundation for an industry-wide reliability database. The database has been designed to make available high quality, valid, and useful data to the hydrocarbon and chemical process industries enabling analyses to support availability, reliability, and equipment design improvements, maintenance strategies, and life cycle cost determination. By design, the database structure contains voluminous detail; however, not all of the detail is required by the user. Each operating company decides the amount of detail input by the user. In fact, the level of detail required for basic analyses is minimal. The detail of the results, however, will be commensurate with the details input to the database.

The foundation for the database was developed from a project directed by the Center for Chemical Process Safety (CCPS). The CCPS has formed a group of sponsor companies to support and direct this effort. The sponsor companies, through representatives on a Steering Committee, are actively participating in the design and implementation of the database.

FIGURE 1.1 ***The role of data in the decision-making process.***

This book, which is expected to be the first in a series, lays out the fundamental concepts of data collection in a way that is intended to provide useful information for decision making as shown in Figure 1.1.

A successful system converts data into information useful for making value added decisions, which can be economic, or safety related in nature. In all cases these translate into business risk as shown in Figure 1.2.

The main objective of this book is to provide the basis for data structure and collection methods to support reliability analyses. In turn, it is hoped that this book will promote the systematic collection and analysis of plant and equipment data. A secondary objective is that this book will serve as a guideline for the CCPS plant and equipment reliability databases and subsequent projects in this field.

Making this happen requires the understanding and integration of several cross-functional disciplines. As such this book is intended for use by

- Maintenance Engineers
- Reliability Engineers
- Design Engineers
- Production Engineers
- Information Technology Software Developers
- Process Safety Risk Analysts

By utilizing basic industry management systems that exist at every facility, it is possible to leverage these systems in a way that provides data in a cost effective manner for further value-added analyses and subsequent decisions. Examples of value-added analyses are best equipment for application, optimal test intervals, knowledge of system or component wear out, optimal maintenance strategy or improved risk analysis. In turn these provide the foundation for achieving improved plant availability and/or reliability and reduced risk.

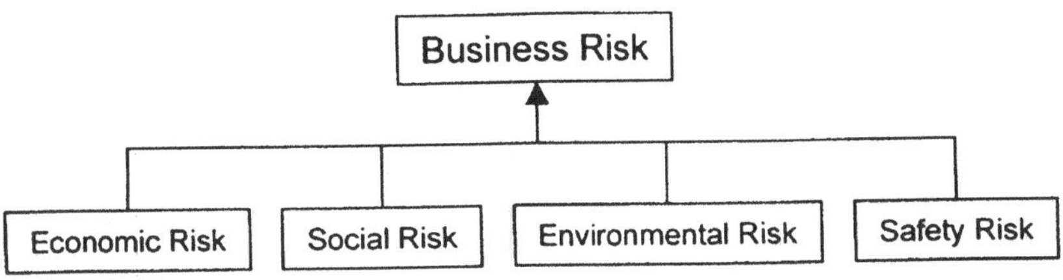

FIGURE 1.2 ***Translating noneconomic risks into business risk***

As the need for more sophisticated analysis increases, so does the need for higher quality, more specific data. This is illustrated by Figure 1.3. Chapters 3 and 4 provide more detail regarding the types of data required for the varying complexity of analyses.

1.2. Taxonomy

An effective taxonomy produces useful data for reliability analyses. The basic building blocks for a reliability analysis fall into two main classes: data regarding specific events occurring for a piece of equipment, and data describing the sample space of equipment. The first class (Event Data) is used as a numerator in the reliability calculation, the second class (Inventory Data) is used in the denominator. Chapter 5 is dedicated to the description of the inherent taxonomy used in the database.

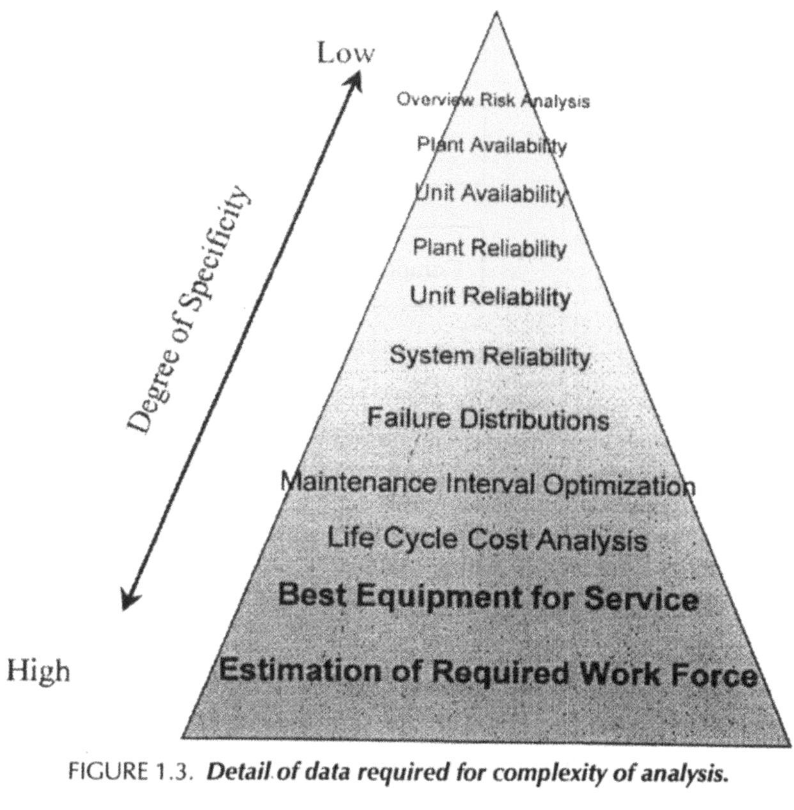

FIGURE 1.3. ***Detail of data required for complexity of analysis.***

The taxonomy and database structures have been developed to facilitate analysis, and are shown in Figure 1.4. A taxonomy such as this will allow sorting and selection of specific types of data, with the types of design, operational, and environmental conditions, which are similar to that of interest in your plant.

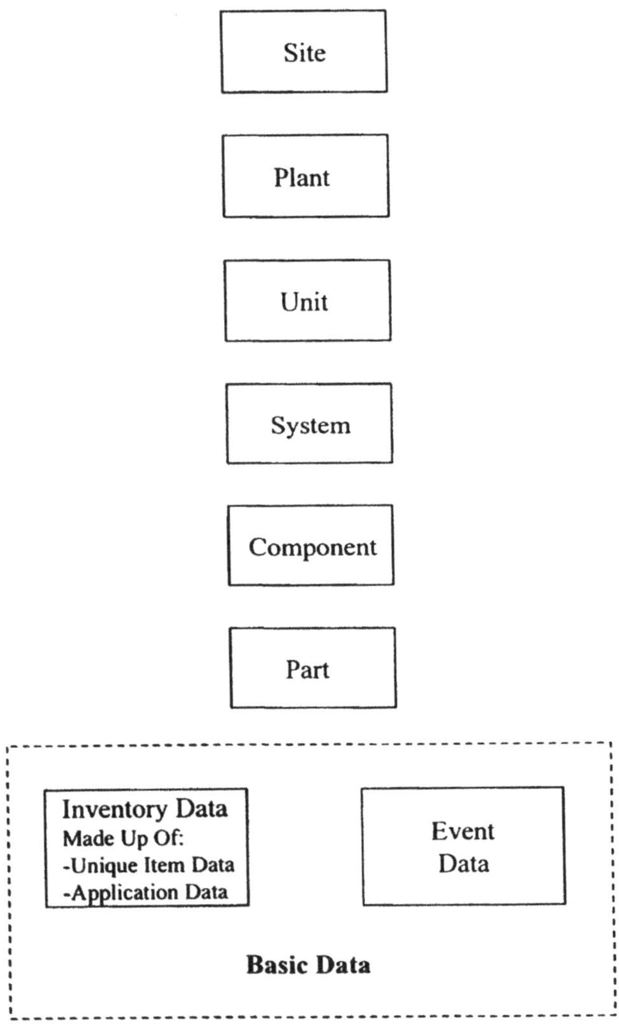

FIGURE 1.4. *CCPS database taxonomy.*

1.3. Data Aggregation/Sharing

CCPS hopes to provide the foundation that allows the hydrocarbon and chemical process industry to establish a standard. This is important because it allows for different systems to become more compatible, increasing the value of industry data. Benefits accrue from being able to aggregate data without degrading its quality. Figure 1.5 shows the CCPS concept, with CCPS acting as an industry clearinghouse.

This concept allows individual plants the ability to measure their performance and to seek improvement in their designs and maintenance strategies. At the company level, it allows the performance comparison of its plants.

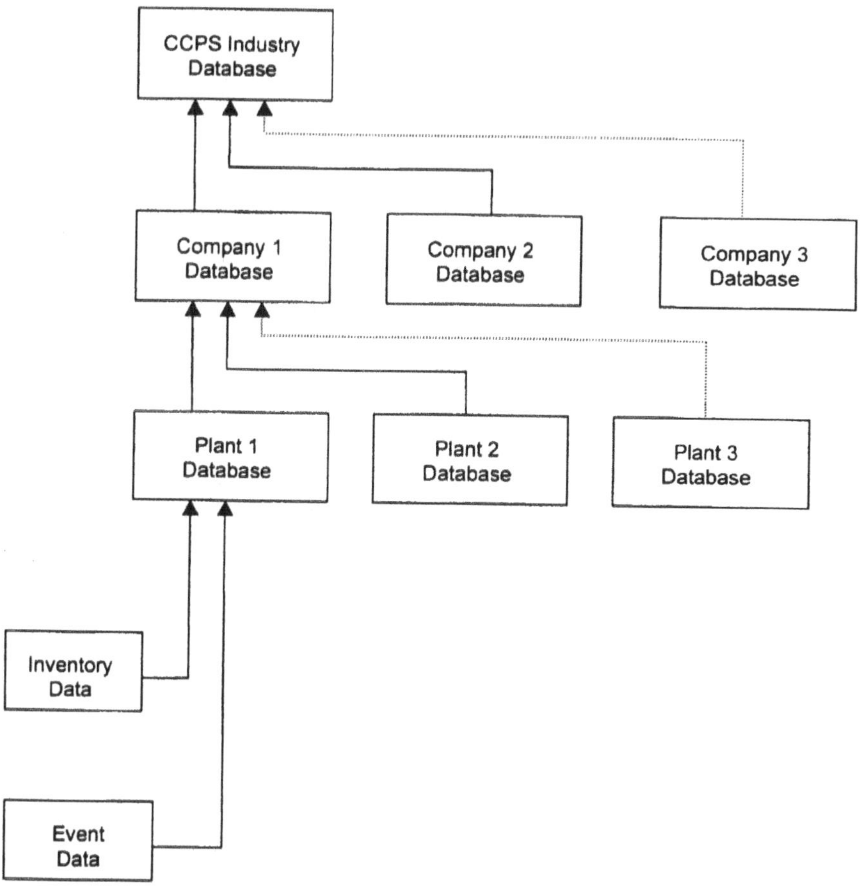

FIGURE 1.5. ***CCPS data aggregation concept.***

Sharing information through CCPS also allows companies to benchmark themselves against industry. It is further a resource of information in areas where they are less expert. The operating concept of CCPS acting as an industry clearinghouse is shown in Figure 1.6. Details of how the data actually flow are provided in Appendix I. After data have been collected and analyzed, it is anticipated that the equipment reliability database will be published and available to the general public.

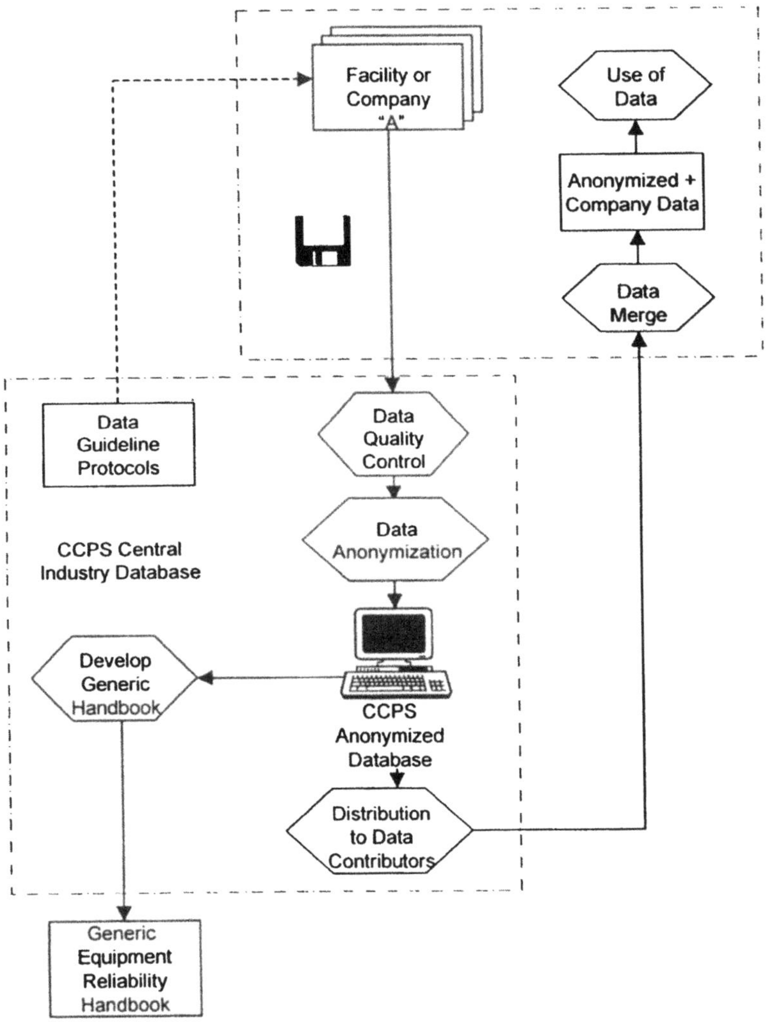

FIGURE 1.6. ***Equipment reliability database operating concept.***

2 Definitions

2.1. Introduction

Within this document we refer many times to a variety of terms that relate to "equipment reliability" and, more specifically, to "equipment failure."

The primary aim of this section is to clarify how we understand this terminology when applied in practice, thus ensuring consistency and commonality across our industries, and more particularly for all subsequent data collection.

For clarity's sake, this chapter is subdivided into two main sections. One section provides a discussion of key terms used throughout the text, terms such as "failures" and "causes." The other section presents other important, though less complex, definitions in a glossary-style format.

2.2. Discussion of Key Reliability Terms

Terminology relating to equipment failures is used throughout the Guideline. Wherever possible, we have adopted readily available explanations that are currently acknowledged by practicing organizations and standards: OREDA and ISO in particular.

In doing so we recognize that this could conflict with views of practitioners who, in the absence of greater clarification, have been using failure terminology with a slightly different meaning to that mentioned herein. In striving for consistency we have undertaken an extensive challenge to be sure that our conclusions meet the best interests of the industry.

Equipment Failures and Related Circumstances

In the practical world, equipment and their components can "fail" under a variety of circumstances. There can be instant failures, partial failures, intermittent failures and a number of other options. This invariably creates differing interpretations amongst engineers as to **what** happened, **why** it happened and **how** it should be categorized, while at the same time wishing to be consistent in communication.

Similarly, the term "failure" is often confused with either or both the terms "fault" and "error."

For the purpose of this project we are adopting the collectively recognized term of "failure mode" as being a key term. As a description, it identifies how we "observe a fault of an item," whether that fault is in its premature stage of failing, has been faulty for a period of time, or has subsequently resulted in the component failing to meet its expectation and has ultimately stopped performing.

We now note that a failure can occur at various stages in an item's life. Hence, we may observe the following:

- A *sudden failure* which, as expected, is a sudden occurrence and causes one or more of the fundamental functions of a component to cease. This type of occurrence is not unusual, but is the one that invariably causes most concern in that it is often unpredictable; although if we could have monitored it we would probably have had time to minimize its impact.
- A *gradual failure* is one in which we are more fortunate to observe a decline in component performance, can probably predict the conclusion and can contain the effect of it actually failing, or can have rectified it before it becomes serious.
- *Partial failures* are associated with defects that do not cause a complete loss of function.
- *Intermittent failures* are situations that result in a loss of performance for a period of time but in which the item reverts to its original working state in due time.

These are all failures and are a result of any combination of more fundamental faults which may relate to design practices, manufacturing procedures, operational styles, maintenance strategies or others. It is the *root cause* of these faults that we strive to resolve.

When the failure actually occurs we need to be able to identify the circumstances that are in being at that time. These are commonly referred to as *"failure descriptors."*

Ultimately, every failure will have an effect on the business/operating environment and this is generally referred to as the *"failure effect."* Clarification may be gained from the two examples that follow.

Practical Examples of Terminology

Example 1. Failed Motor Bearing

Putting this into reality, we might consider the simple example of an electric motor that has a "fault." The detailed nature of that fault, in this example, is not known at the time the plant operator observes that something is not correct. However he registers the defect (fault) in his computerized maintenance management system as work to be investigated. For the time being we shall assume that it is a *sudden failure.*

The plant operator's *observation* (or description) is referred to as the **failure mode**: for example, the motor failed to start on request (Fails to Start). It is the description or observation of the fault that is happening as viewed from the outside. Very often there are designated acronyms to facilitate coding of such information, for example, FTS.

This *failure mode* ("failed to start" in this case) could equally apply to other items of equipment.

(Other failure modes might be, "valve closing too fast," "pipeline plugged," "display signal not functioning," etc.)

Who it is that registers these details into the system and ensures information consistency is left to the discretion of the owner, but, as was mentioned earlier, quality of information should not be overlooked.

In our example, investigations by the maintenance organization subsequently show that the apparent *observed cause* of the failure was a bearing problem. This is typically referred to as the **"failure descriptor"**; that is, it describes in brief terms what was found to have *actually failed* as against what was observed from an operational viewpoint.

Let us assume, in this case, that the **"failure cause"** associated with this example was that the wrong type of bearing had previously been substituted for the correct one, at some time in the past. "Failure cause" thus relates to the information that is required to avoid the failure recurring. These may relate to design, manufacture, misuse, aging, or other factors.

The **"root cause"** would be, for example, that clear maintenance instructions were not applied to ensure that alternatives were installed without thorough consideration of the consequences.

We hope that you can visualize that we are first observing a problem (failure mode) and then identifying the defect (failure descriptor), the reason for the failure (failure cause), and the fundamental problem (root cause).

The **"failure effects"** associated with this problem are *the consequences that apply.* In our example we will assume that the alternate machine was commissioned, and minimal loss of production occurred.

Example 2. Failed Lube Oil Cooler of a Centrifugal Compressor
In this example we will assume the cooler tubes were heavily corroded resulting in water contamination the lubrication system.

The "failure mode" was reported by the routine oil-sampling team (or laboratory, or ferrography group, or condition-monitoring group). The failure is thus examined, and the "failure descriptor" identifies (after examination) that it was the tubing in the lube oil cooler which had corroded (it could have been a joint failure or a number of other alternatives). The "failure cause" culminated from the eventual corrosion of some oil cooler tubes, and the "root cause" was identified as being unsuitable material selection for the duty. The "failure effect" was that no damage occurred to the compressor bearings. The lube oil system had to be drained and replaced; the machine was idle for a period of time; and production was lost.

This example *could* have been a compound situation where, if routine lube oil sampling was not applied, the bearings could have failed, the rotor could have been damaged, and many other problems would have been experienced.

2.3. Glossary of Terms

The following glossary provides definitions for terms commonly used throughout these Guidelines.

Term	Definition
Aggregation	The pooling of compatible data from disparate sources to create a larger population of raw data for analysis.
Application	Information described by the process parameters associated with a specific location. Equipment system which is linked to a process unit by a specific tag number.
Availability	The ability of an item to be in a state to perform a required function under conditions at a given instant of time or over a given time interval, assuming that the required external resources are provided. For clarity, refer to Chapter 3.
Boundary	Specification of the interface between an item and its surroundings.

Term	Definition
Censored data	Data that are not complete (where the ideal testing scenario does not occur).
• Right-censored data	Data collected from the time at which an item is put on test at time 0 and is removed during the test period while the item is still working.
• Left-censored data	Data collected from the time at which an item is put on test at time 0 and found to be functioning, inspected again after an interval of time and found to have failed.
• Interval-censored data	Data collected from the time at which an item is put on test at time 0, inspected some time later and found to be have survived, inspected again and found to have failed.
• Singly censored data	Right or left censored data.
• Multiply censored data	Data in which any or all (right, left, interval) censoring types may be present
Complete data	Data collected from the time at which an item is put on test or in service at time 0 and is monitored continuously until it fails at time t.
Component	An equipment item composed of one or more parts that are capable of being repaired.
Corrective maintenance	The maintenance carried out after fault recognition and intended to put an item into a state in which it can perform a required function.
Database	A repository for equipment reliability information categorized to facilitate normal relational database management functions (e.g., sorting, filtering).
Database administrator	An individual responsible for the administration of the CCPS Process Equipment Reliability Database.
Database contributor	An operating entity that provides valid data to the CCPS Process Equipment Reliability Database. This is considered to be a corporation, although one or more people may represent the interface to the database.

Term	Definition
Database participant	A member of the CCPS Database project, that may or may not be a database contributor.
Data subscriber	The point source contact between the database contributor and the CCPS database administrator.
Demand	Activation of the function (includes both operational and test activation).
Down state	A state of an item characterized either by a fault, or by a possible inability to perform a required function (e.g., during preventative maintenance).
Downtime	The time interval during which an item is in a down state.
Equipment group	A convenient means to organize similar types of systems (e.g., pumps, compressors, turbines are part of the equipment group "Rotating Machinery").
Error	A discrepancy between a computed, observed or measured value or condition and the true, specified or theoretically correct value or condition.
Failure	The termination of the ability of an item to perform a required function.
• Complete failure	A failure that is a complete loss of a required function.
• Extended failure	A failure that results in a lack of some function that will continue until some part of the item is replaced or repaired.
• Intermittent failure	A failure that results in a lack of some function for a very short period of time. The item will revert to its operational state without repair.
• Partial failure	A failure that leads to a loss of some function but does not cause a complete loss of a required function.
• Sudden failure	A failure that could not be forecast by prior testing or examination.

Term	Definition
• Gradual failure	A failure that could be forecast by testing or examination.
Failure cause	The circumstances during design, manufacture or use which have led to a failure.
Failure descriptor	The apparent, observed cause of failure.
Failure mechanism	The physical, chemical, or other process or combination of processes that has led to a failure.
Failure mode	The observed manner of failure. The failure modes describe the loss of required system function(s) that result from failures.
Fault	The state of an item characterized by inability to perform a required function, excluding the inability during preventive maintenance or other planned actions, or due to lack of external resources.
Function	The normal or characteristic actions of an item.
Industry	A distinct grouping of facilities that manufactures, processes or delivers product(s) having similar characteristics or applications.
Inventory	Equipment population information that is made up of application data and data that are specific to an equipment item regardless of application.
Item	Any part, component, device, subsystem, functional unit, equipment or system that can be individually considered.
Maintenance	The combination of all technical and administrative actions, including supervision actions, intended to retain an item in, or restore it to, a state in which it can perform a required function.
Operating mode	
• Alternating/ Cyclical	A mode in which an item, in normal operation, repeatedly and predictably changes its operating conditions to perform its required functions, whereupon it returns to its original condition automatically.

Term	Definition
• Batch	A mode in which an item, in normal operation, performs or permits a sequence of operations to take place and which, after completion of the operations, is returned to its original state either automatically or through manual intervention.
• Running/ Continuous	A mode in which an item is performing a single active operation at specific operating conditions.
• Standby	A mode in which an item is not actively running, but can be made to actively run on demand.
Operating state	
• Normal	The state in which an item is performing its usual, nontransitional required function. This state can include any of the operating modes listed above.
• Shutdown	The state in which an item is in a transitional mode from being in its normal operating state to being taken off-line.
• Startup	The state in which an item is in a transitional mode from being off-line to being in its normal operating state.
Operating time	The time interval during which an item is in an operating state.
Part	An item that is not subject to disassembly or repair.
Plant	A process facility, at a discrete location, that manufactures a product(s).
Preventive maintenance	The maintenance carried out at predetermined intervals or according to prescribed criteria and intended to reduce the probability of failure or the degradation of the functioning of an item.
Redundancy	In an item, the existence of more than one means for performing a required function.
Reliability	The probability of an item to perform a required function, under given conditions, for a given time interval.

Term	Definition
Repairable item	Durable item determined by application of engineering, economic, and other factors to be restorable to serviceable condition if it fails.
Nonrepairable item	An item that is replaced, not repaired, upon failure.
Root cause	Fundamental cause of failure. The cause(s) for which remedial actions can be decided upon.
Site	A process facility, at a discrete location, consisting of one or more plants.
Subordinate system	A system that provides a function exclusively to another system.
Subscriber	A person or job title that is registered and qualified to submit data to the CCPS Equipment Relibility Database in accordance with an approved data contributor quality plan.
Surveillance period	The interval of time between the start and end of data collection.
System	A group of subordinate system(s) or component(s) or part(s) connected or associated in a fixed configuration to perform a specified function(s), as described by a database boundary diagram.
Taxonomy	A systematic classification of items into generic groups based on factors possibly common to several of the items (e.g., functional type, medium handled).
Time to repair—active	The cumulative time actually spent repairing an item and ensuring it is ready for service (see Figure 3.2 in Chapter 3).
Time to repair—total	The time interval from occurrence of failure until the item is ready to be returned to service (see Figure 3.2 in Chapter 3).
Time to restore	The time interval from occurrence of failure until the application is restored, inclusive.

Term	Definition
Time to failure	The time interval from the point at which an item is put on test or in service until failure occurs.
Time to shutdown	The time interval from the point at which an item is put on test or in service until it stops running.
Unit	A collection of interrelated systems, the purpose of which in normal operation is to process one or more inlet feedstocks or intermediates by either physical or chemical means, to produce outlet intermediates or products having one or more of its phase, composition, temperature and/or pressure changed and delivered.

3 Methods of Analysis

3.1. Introduction

This chapter describes practical methods for estimating reliability parameters for items included in the CCPS Equipment Reliability Database taxonomy. We focus on the types of analyses that process reliability engineers are most likely to encounter. Formal statistical details are left to the references.

The types of analyses presented herein are broken down into two broad classifications: those based on sample statistics and those based on inferences from statistical distributions. The distinction between the two classes is subtle, but important to the reliability engineer. Reliability parameters based on sample statistics are fairly simple in form, but limited in usefulness. Although characteristics of distributions are more difficult to derive, they provide much more relevant information to the reliability engineer. Experience indicates that some probability distributions are more representative and more effective in certain situations than others.

The purpose of this chapter is to provide practitioners with a brief source of reference material for analyzing failure data, and to illustrate the concepts behind the calculations performed by the CCPS Database software. Further information about the specific algorithms used in the software can be found in Appendix I. This chapter covers basic concepts in data analysis, along with an overview of some of the many statistical distributions available. It concludes with a discussion of hypothesis testing and a brief survey of more advanced topics on data analysis.

The chapter is not intended to be a comprehensive source of information on the topic of data analysis. A list of references is provided at the end of the chapter to provide the practitioner with more details on the subject.

3.2. Basic Concepts of Data Analysis

3.2.1. Failure Data

For most equipment items in service, it is natural to observe different times-to-failure when those items are placed in operation simultaneously and under identical conditions. This is because we can expect a certain amount of random behavior in engineered equipment performance. The variations in performance can become greater when new items and old items are compared. Therefore, the value associated with the time-to- failure is not fixed; rather, it follows a distribution across a range of values.

Much of the data analysis work revolves around the correct classification of the time-to-failure data and the estimation of the statistical parameters for known distributions in order to represent the broader population and estimate the equipment reliability.

The process of extrapolating failure data into reliability results can be quite complex. Much of the work is done by trial-and-error, and is therefore difficult to proceduralize. Presentation of some basic concepts is therefore paramount. Worked examples will follow presented concepts in order to demonstrate how they may be put into practice.

3.2.2. Need for Correct Failure Modes

A failure mode, as defined in Chapter 2, is a description of how a failure is observed. This observation comes about by observing the affect on an equipment item's ability to perform. Failure data will vary according to failure mode. Therefore a critical step in analyzing data is to identify the failure modes correctly. If failure modes are incorrectly assigned, data variance can be very large, resulting in poor-quality reliability estimation.

A given cause could lead to more than one mode and a mode could have more than one cause. Therefore, it may not be possible to gather information useful to prevent failure recurrence through grouping data by mode alone. One tends to focus on the mode because it helps highlight the *effect* of the failure. For predictive analysis, however, the functions are linked to failure cause/mechanism, irrespective of effect.

3.2.3. Types of Systems—Repairable or Nonrepairable

The issue of repairability can have a significant effect on the process used for data analysis. For the CCPS equipment reliability data project, most *systems* and *components* are repairable, whereas most *parts* are not. Keep in mind that failure detection is part of the repair process and that undetected failures remain unrepaired.

Repairable items can experience a large number of failures over their lives since they are repaired and restored to operation after a failure. Instead of a single time-to-failure measurement on an item, we have a sequence of times between failures. The models and methods for analyzing these kinds of items are different from those for analyzing data from nonrepairable items.

3.2.4. Reliability versus Availability

Repairability of an item affects two main reliability parameters: availability and reliability. Availability can be considered as the fraction of time that an item or group of items is in a state of being able to perform its intended function, as calculated by past performance. Reliability, on the other hand, is the probability that an item has survived to a certain time without failure. If an item is repairable, it may experience a prior failure but still be available later at a given time. Under these conditions, however, it may not be considered "reliable" (bearing in mind that being "reliable" is relative to what one considers to be acceptable performance). Methods for estimating these two parameters vary depending on whether or not the item is repairable.

Although the *reliability* of an item is of fundamental interest, ultimately it is the availability of the overall plant that is critical to the economic success of the facility. There are many methods, such as Monte Carlo, that can model the effect of individual system behaviors on the overall plant availability. Discussion of these is beyond the scope of this book. However, the CCPS Database does collect and allow retrieval of plant availability information.

3.2.5. Types of Data—Censoring

The ideal scenario for failure testing is one in which the following are true:

- The item is placed into service at the very start of the test.
- A failure occurs and is detected during the "window" of time the test is conducted.
- Repairs return the equipment to as good as new status.
- There are no changes in the operating envelope.

Unfortunately, this is not often the case in the process industry. If the ideal testing scenario does not occur, the data are said to be *censored*. The CCPS Database calculations assume that the last two bullets are true, although the capability is also provided to remove data that do not conform to the analysts' requirements. Censoring of data, as described in the first two bullets, affects the precision with which the times to failure have been measured. There are four basic types of data: complete, right-censored, left-censored, and interval-censored. These are depicted in the diagrams below.

3.2.6. Definitions

Exact failure or *complete data* occurs when an item is put on test at time 0 and is monitored continuously until it fails at time *t*. Time is assumed to be measured without error. In this case we know the failure time of the item exactly. An example is that of a manufacturer putting 1000 bearings on test and continuing the test until all of the bearings have failed. The diagram below illustrates a time-line for complete data.

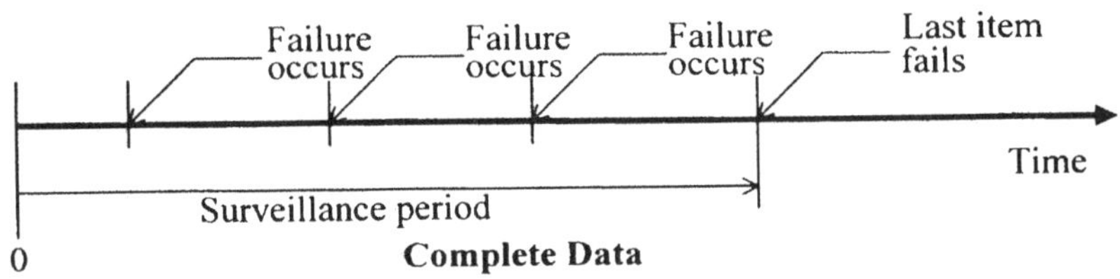

Complete Data

For *right-censored data*, an item is put on test at time 0 and is removed during the test period while the item is still working. In this case we only know that the time to failure is greater than the time to removal. An example is a manufacturing "burn-in" of electronic chips.

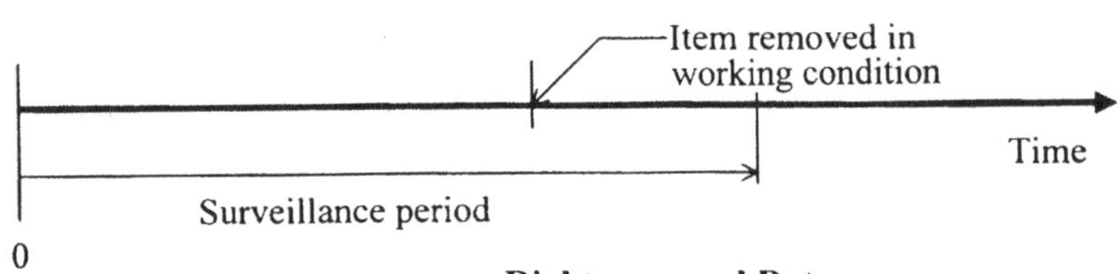

Right-censored Data

Data are considered *left-censored* if the item is put on test at time 0, but the exact failure time of the item is not known. This could occur if the item is put on test and not monitored until some time later when failure is detected. In this case we only know that the time to failure is less than the time of detection. An example is a failure of a relief valve that is detected during a proof test.

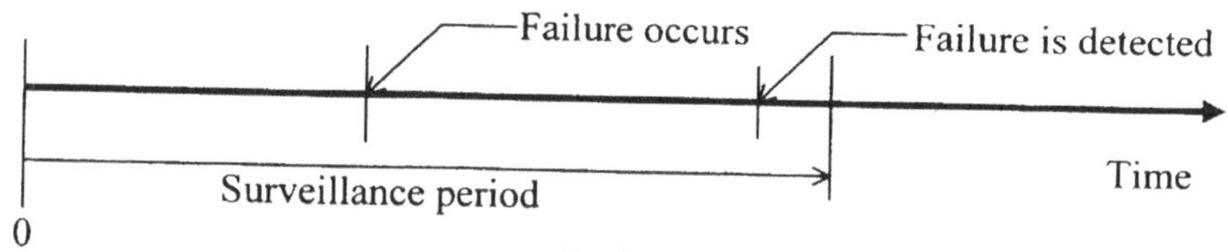

Left-censored Data

Interval-censored data occur when an item is put on test at time 0, inspected some time later and found to be functioning, inspected again after an interval of time and found to have failed. In this case all we know is that the time to failure falls in the interval between the times of the two inspections. Examples include failures discovered during a periodic inspection program, when the true failure time is not known.

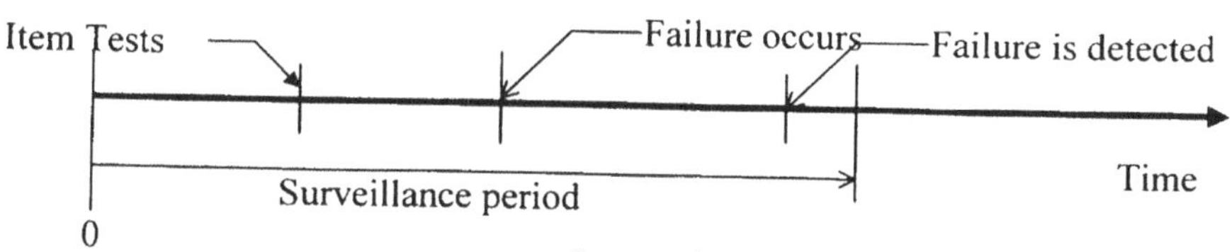

Interval-censored Data

3.2.7. Dealing with Censored Data

It is often necessary to estimate the failure rate for a population from a sample data set of times-to-failure. Some of these data points may be incomplete, for instance, when an observation ends before all the items have failed. This is generally the case with industry data. In actuality, the true value of the failure rate for a particular failure event out of a sample of items is itself a random variable for data that are independently generated and identically distributed. There are trend tests to check this and the reader is referred to the references for more information.

A common method of dealing with censored data is to arrange all the data in time order and assign an order number to each failure for the particular mode being studied. In general, the order numbers of the failures following the first censoring will no longer be integers, but will take fractional values to allow for the censored item. The general formula for the next increment, that is, as soon as the first censoring is encountered is

$$\text{New Increment} = \frac{(n-1) - (\text{Previous Rank Order Number})}{1 + (\text{Number of Items Beyond Present Censored Item})}$$

where n is the number of items in the sample.

This increment is added to the previous rank order number until another censored item is encountered, at which time a new increment is calculated. In general, the next failure will have a rank order number equal to the preceding number plus the new increment. In order to check that the calculations are carried out correctly, the following relationship should be verified:

$$\text{Last Increment Calculated} + \text{Last Rank Order Number} = n + 1$$

Median ranks and confidence limits can then be determined from the median rank tables using the sample size column and interpolating fractional order numbers. More discussions pertaining to ways of dealing with censored data are provided in an example which follows, and in Chapter 4.

Although simple analytical procedures for rigorously analyzing interval-censored data do not exist, special software systems capable of analyzing such data are available. Descriptions of the characteristics of these systems are beyond the scope of this chapter and may be found in detail in technical references.

3.2.8. Common Cause Failures

Some failures of a given equipment item may be related. For example, plugging in a relief valve may lead to a failure to open at the set pressure, and may also lead to the material not being relieved at the specified rate. On a larger scale, a single power outage may result in the shutdown of many pieces of equipment. The CCPS Database software makes no distinctions between the "common cause" failures at the raw data level, but allows the data analyst to segregate such events in the data retrieval process to avoid double-counting.

3.2.9. Predictive versus Descriptive Methods

This chapter deals exclusively with predictive analysis methods. These are not the only methods available in reliability, although they are the most powerful if data in the required amount of detail is produced. Predictive methods require a lot of data which must be independent and identically distributed (IID). In some cases, this type of data is just not available in the volumes required.

Descriptive methods are used for monitoring purposes and do not give clear information on the causes of failure. However, they can be used to check estimates and to detect significant changes in the component reliability performance. They are techniques borrowed from production statistics and can often be applied even when the volume of data precludes the more in-depth predictive approach.

In predictive analysis the main variable is time to failure (TTF). By fitting TTF to a function we can derive information on availability, MTTF, the percentage of items which can be expected to have failed by a given time (not the same for different mechanisms of failure), and the nature of the failure. This information is very useful to maintenance engineers to identify the most suitable action for failure prevention.

In descriptive analysis the main concern is spotting unacceptable deviations from average performance, identifying trends and, if the behavior is stable, deriving estimates for the average number of failures we can expect in normal operation. This is of great importance to the maintenance manager for the procurement of spares and maintenance researching, feedback to designers on performance against production targets, etc.

In practice, unless we run a test plant, pure TTF data are rarely available. Therefore, data will be derived mainly from repairable plant items. However, in the predictive approach the data are treated as if they represented individual TTF of nonrepairable items. This approximation is legitimate if the data are IID. By contrast, in descriptive analysis the data are treated as counts of failures associated with repairable items—no assumptions need to be made on the underlying distribution or the independence of subsequent events.

3.3. Data Manipulation Examples

3.3.1. Methods of Analysis

Once the necessary data are available, the mathematical analysis can proceed. This section identifies the various reliability parameters and outlines the fundamental mathematical relationships among them. Numerical examples are also provided to illustrate their application in practice. Two cases are considered: nonrepairable and repairable items.

Whether the items being studied are repairable or not, there are several important assumptions that govern the analysis:

1. At any given time, an item is either functioning or failed.
2. A new item jumps into a normal state, then fails and is transformed to a failed state.
3. A failed state continues indefinitely if the item is nonrepairable.
4. A repairable item, after repair, goes back to a normal state.
5. The change between states is instantaneous.

3.3.1.1. Nonrepairable Items (Repair-to-Failure Process)

Failure in this case means loss of function. Consider the following new symbols:

$f(t)$ = Probability density function of $F(t)$.
n = Total number of items in sample.
$Nf(t)$ = Number of failures occurring before time t.
$h(t)$ = Instantaneous failure rate, it is the probability that an item experiences a failure per unit time at time t, given that the component was normal (repaired) at time zero and has survived to time t. Also known as hazard rate.
$H(t)$ = Cumulative hazard value: $H(t) = \int h(t)\,dt$
$L(t)$ = Number of items functioning or surviving at time t.
$r(t)$ = Repair rate, it is the probability that an item is repaired per unit time at time t, given that the item failed at time zero and has been failed to time t.
Δ = Increment of time measured from one particular instant to the following moment of interest.

Typical sample data would provide information pertaining to the values of t and $L(t)$ throughout the entire period of data surveillance. The reliability $R(t)$ is the probability of survival up to and including time t; or

$$r(t) \frac{\text{Number of Items Surviving at Time } t}{\text{Total Sample Size at Time 0}} \tag{3.1}$$

The unreliability $F(t)$ is the probability of failure up to time t, or

$$F(t) \frac{\text{Number of Failed Items at Time } t}{\text{Total Sample Size at Time 0}} \tag{3.2}$$

From probability theory, $R(t)$ and $F(t)$ are related by the mathematical relationship:

$$R(t) + F(t) = 1 \tag{3.3}$$

The distribution function $F(t)$ may be used to determine the proportion of items expected to fail during a certain time interval $t_2 - t_1$, as:

$$\frac{Nf(t_2) - Nf(t_1)}{n} = F(t_2) - F(t_1) \tag{3.4}$$

Parameters of interest to the reliability engineer may include $R(t)$, $h(t)$, $r(t)$, the mean time to failure (MTTF), and the mean time to repair (MTTR). Given that values of t and $L(t)$ are known, and that t values range from zero to the entire duration of the test, a simple systematic procedure similar to that illustrated in the example shown below may be employed. Again, note that "time" in this context refers to time since either "item birth" or since the last repair, where the repair made the item "as good as new."

The reliability parameters are calculated from the following relationships in a systematic manner:

Reliability $R(t)$:

$$R(t)\frac{L(t)}{n} \tag{3.5}$$

Instantaneous failure rate $h(t)$:

$$h(t) = \frac{f(t)}{R(t)} = \frac{f(t)}{1 - F(t)} \tag{3.6}$$

where the failure density $f(t)$ is obtained from

$$f(t) = \frac{Nf(t + \Delta) - Nf(t)}{\Delta n} \tag{3.7}$$

Repair rate $r(t)$:

For nonrepairable items the repair rate is constant and equal to zero.

The parameters above are not calculated in the CCPS Database software. These can be derived, however, by exporting data from the Database and applying either manual methods or using software for this purpose.

Two values of particular interest to the reliability engineer, which can be deduced from the above equations are the mean time to failure MTTF, defined as the expected value of the time-to-failure, and expressed by

$$\text{MTTF} = \int_0^\infty tf(t)\,dt \tag{3.8}$$

and the mean time to repair, MTTR, defined as the expected value of the time-to-repair, TTR, and expressed by

$$\text{MTTR} = \int_0^\infty tg(t)\,dt \tag{3.9}$$

Note that these values are means, not medians. Thus, if these functions are skewed in any way, something other than 50% of items will have failed at the "mean" time (e.g. ~60% for an exponential distribution.) The MTTF and MTTR can be calculated rigorously using a hazard plot approach as shown below, or via a numerical integration illustrated in Chapter 4. Alternately, they can be estimated assuming that failure and repair times follow an exponential distribution, as is assumed in the CCPS Database software. The assumption of exponential distribution, and alternative distributions, are discussed later in this chapter. The example below illustrates the comparison between the hazard plot and exponential distribution approaches.

Example 1 (Complete Data)
Table 3.1 shows failure data for 50 transistors, all of which failed during the test period. Calculate the unreliability $F(t)$, the failure rate $r(t)$, the failure density $f(t)$, and the MTTF.

The problem will be solved step by step to illustrate the methodology outlined in Section 3.3. The following steps will be followed.

Step 1. *What type of failure data are these?* It is important to assess what type of failure data will be available.

Step 2. *Preliminary graphical analysis.* While it is common practice to assume that the time between failures is distributed according to an exponential distribution, this could be a risky assumption, especially if the instantaneous failure rate, $r(t)$, is *not* a constant. A simple graphical analysis could shed some light on the validity of this assumption. There are a variety of nonparametric procedures that could be used—Kaplan-Meier, actuarial, etc.

Step 3. *Numerical computations.* Once the candidate distributions have been identified, proceed with the numerical computations. It is possible that no distribution represents the data appropriately, in which case an empirical approach is recommended.

Solution:
Step 1. This is a case of complete data, since every component has failed during the observation period and all failure times are known. Obviously, this is the simplest case.

Step 2. Graphical analysis. A procedure recommended by Nelson (see Reference 5 for a more complete discussion) and slightly modified for this example will be adopted. The procedure works as follows:

1. Order the n failure times from smallest to largest. Label the times in reverse order, that is assign the number one to the largest item.

TABLE 3.1
Failure Data for Transistors

Device Number	Failure Time	Device Number	Failure Time	Device Number	Failure Time
1	9	21	275	41	969
2	14	22	302	42	1084
3	25	23	358	43	1119
4	30	24	380	44	1132
5	59	25	385	45	1199
6	122	26	392	46	1440
7	128	27	426	47	1599
8	138	28	426	48	1648
9	144	29	451	49	1800
10	146	30	453	50	1965
11	148	31	600		
12	151	32	641		
13	177	33	681		
14	183	34	712		
15	202	35	733		
16	206	36	750		
17	246	37	864		
18	259	38	880		
19	267	39	898		
20	275	40	950		

2. Compute an estimate of the hazard value, $r(t)$, for each failure as $1/k$, where k is its reverse rank.
3. Compute the cumulative hazard value, $H(t)$, for each failure as the cumulative sum of the hazard values computed in the previous step. See Table 3.2 for results.
4. Choose a candidate distribution. The typical initial choice is the exponential distribution (the CCPS Database automatically assumes this type of distribution). Do the following transformations:
   Since,

$$F(t) = 1 - e^{-H(t)},$$

then

$$1 - F(t) = e^{-H(t)}.$$

Taking logs on both sides:

$$\ln[1 - F(t)] = -H(t)$$

Let $y = \ln[1 - F(t)]$, giving rise to the equation $y = -H(t)$. If the exponential distribution is a good choice, $H(t) = \lambda t$, where λ is a constant failure rate assuming an exponential distribution. Hence, $y = -\lambda t$. That is, the equation of a straight line that goes through the origin. $1/\lambda$ represents the MTTF for an exponential distribution.

5. A simple regression package can be used to fit $H(t)$ versus t. For the exponential distribution to be a good assumption, the linear fit should be good and the intercept term not significant, i.e., zero.

Running a regression program on $H(t)$ versus t yields the following basic results:

$R^2 = 0.985$

	Coefficients	Standard Error	*P*-value	Lower 95%	Upper 95%
Intercept	–0.069330425	0.035402755	0.05601595	–0.140512393	0.001851543
Time	0.00187991	4.67433E-05	1.2534E-38	0.001785926	0.001973894

The basic hypothesis regarding the intercept is that its true value is zero. Since the 95% confidence interval (–0.1405 to 0.0018) includes zero, the hypothesis that the true value of the intercept is equal to zero cannot be rejected. This means the regression line may be assumed to go through the origin as expected. In addition, the estimate of $1/\lambda$ (MTTF) is given by the "time" coefficient, i.e., $\lambda = 0.00187991$, which implies MTTF = 531.94. This is, of course, a rough estimate.

Now that it has been found that the exponential distribution is likely to be valid, a better estimate of MTTF is obtained by taking the sum of the failure times divided by the sample size:

$$\text{MTTF} = \frac{\sum \text{Failure Times}}{\text{Sample Size}} = \frac{28{,}441}{50} = 568.82$$

While it is tempting to compute MTTF directly using the above formula, it is important to emphasize that the equivalence of MTTF to $1/\lambda$ is not valid if the underlying distribution is not exponential.

As a general comment, a statistical package like SAS would produce the following estimates, based on standard regression analysis:

TABLE 3.2
Computation of r(t) and H(t)

	Fail Time	Rank	$r(t)$	$H(t)$		Fail Time	Rank	$r(t)$	$H(t)$
1	9	50	2.00	2.00	26	392	25	4.00	72.32
2	14	49	2.04	4.04	27	426	24	4.17	76.49
3	25	48	2.08	6.12	28	426	23	4.35	80.84
4	30	47	2.13	8.25	29	451	22	4.55	85.38
5	59	46	2.17	10.43	30	453	21	4.76	90.15
6	122	45	2.22	12.65	31	600	20	5.00	95.15
7	128	44	2.27	14.92	32	641	19	5.26	100.41
8	138	43	2.33	17.25	33	681	18	5.56	105.97
9	144	42	2.38	19.63	34	712	17	5.88	111.85
10	146	41	2.44	22.07	35	733	16	6.25	118.10
11	148	40	2.50	24.57	36	750	15	6.67	124.76
12	151	39	2.56	27.13	37	864	14	7.14	131.91
13	177	38	2.63	29.76	38	880	13	7.69	139.60
14	183	37	2.70	32.46	39	898	12	8.33	147.93
15	202	36	2.78	35.24	40	950	11	9.09	157.02
16	206	35	2.86	38.10	41	969	10	10.00	167.02
17	246	34	2.94	41.04	42	1084	9	11.11	178.13
18	259	33	3.03	44.07	43	1119	8	12.50	190.63
19	267	32	3.13	47.20	44	1132	7	14.29	204.92
20	275	31	3.23	50.42	45	1199	6	16.67	221.59
21	275	30	3.33	53.76	46	1440	5	20.00	241.59
22	302	29	3.45	57.20	47	1599	4	25.00	266.59
23	358	28	3.57	60.77	48	1648	3	33.33	299.92
24	380	27	3.70	64.48	49	1800	2	50.00	349.92
25	385	26	3.85	68.32	50	1965	1	100.00	449.92

Exponential Parameter Estimates				
			Asymptotic Normal 95% Confidence Limits	
Parameter	Estimate	Std Err	Lower	Upper
Scale	568.8196	80.4432	431.1184	750.5032
Shape	1.0000	0.0000	1.0000	1.0000

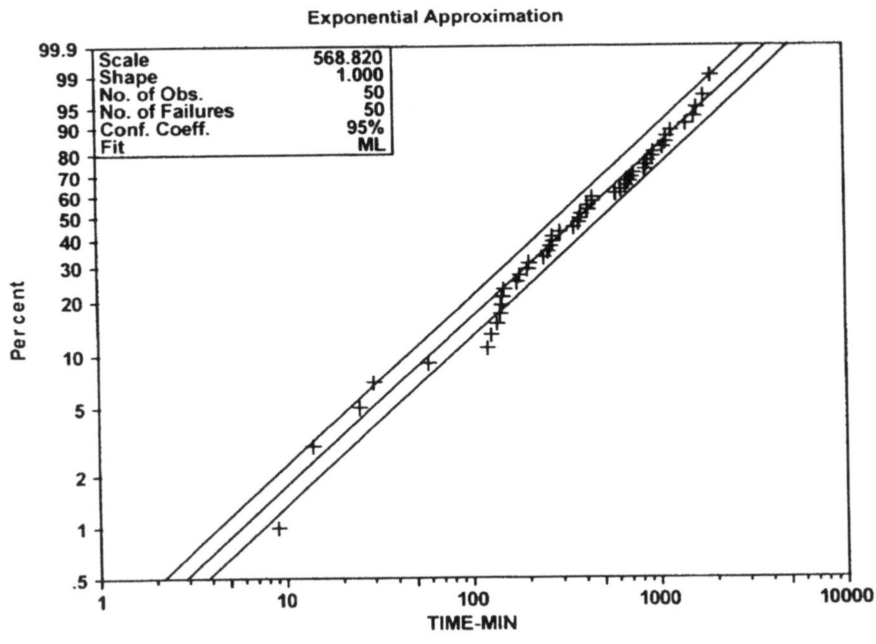

FIGURE 3.1. ***Output from SAS—Exponential distribution fit.***

The scale parameter is the MTTF.

This procedure can be easily extended to the Weibull distribution (see Chapter 4). There are a number of commercial packages available for analyzing distributions. Even widespread software such as Microsoft EXCEL contains statistical functions of limited utility.

Example 2 (Left-Censored Data)

This example is typical of equipment that is in standby mode, and thus may have failures that are undiscovered until a demand or test is made.

Since the times of failures are not known, it is assumed that the failures occur in the midpoint of the time interval from when the item was put in service and when it was removed for testing. It is also assumed that the exponential distribution applies. Table 3.4 illustrates the simple calculation that results.

Note that the precision of this analysis depends on the relative width of the interval between tests with respect to the typical time scale of failure phenomena. In this case the midpoint assumption is not so damaging. However, if the failure time scale is much shorter than the test interval we cannot hope to derive much useful information from the procedure. This can happen in real life, for example when a component suffers from early failures.

TABLE 3.3
Test Data for Relief Valves

Proof Test Time (years)	Success/Failure?
0.5	Success
1.0	Success
1.0	Failure
1.0	Success
1.5	Success
1.5	Failure
1.5	Success
2.0	Success
2.0	Success
2.0	Success
2.5	Failure
2.5	Failure
2.5	Failure
2.5	Success
3.0	Failure
3.0	Success
3.0	Failure

TABLE 3.4
Calculation of MTTF for Relief Valves

Proof Test Time (years)	Success/Failure?	Estimated Time to Failure/Success (years)
0.5	Success	0.5
1.0	Success	1.0
1.0	Failure	0.5
1.0	Success	1.0
1.5	Success	1.5
1.5	Failure	0.75
1.5	Success	1.5
2.0	Success	2.0
2.0	Success	2.0
2.0	Success	2.0
2.5	Failure	1.25
2.5	Failure	1.25
2.5	Failure	1.25
2.5	Success	2.5
3.0	Failure	1.5
3.0	Success	3.0
3.0	Failure	1.5
	Total Time in Service (years)	25.0
	Number of Failures	7
	MTTF (years)	3.57

3.3.1.2. Repairable Items (Repair–Failure–Repair Process)

Repairable items or components experience repetitions of the repair-to-failure and failure-to-repair processes. The reliability characteristics of such items can be calculated by considering each item to be a sample from a population of similar items undergoing identical use. This is a great simplification of the complexity underlying repairable system failure analysis. This assumption has the advantage of transforming the problem at hand into an equivalent, much easier, problem where we have a larger sample (larger than the actual sample size) consisting of nonrepairable items. The repairability property of an item does not only affect its reliability but also has an impact on its availability.

Availability $\bar{A}$

One of the simplest performance parameters for a process plant, unit operation, or equipment system is its availability. Availability is the fraction of time an item (site, plant, unit, etc.) is in working order. It is generally associated with a complex system or groups of systems that are repairable. The general expression for availability based on an average over a time ensemble is given by

$$\bar{A} = \frac{\sum_{i=1}^{n} \mathrm{TTF}_i}{\sum_{i=1}^{n} (\mathrm{TTF}_i + \mathrm{TTR}_i)} \tag{3.10a}$$

where TTF_i is the span of time from repair to first subsequent failure for item i, n is the total number of items at time 0, and TTR_i is the length of time from failure to first repair of item i. This can also be expressed as

$$\bar{A} \frac{\mathrm{MTTF}}{\mathrm{MTTF} + \mathrm{MTTR}} \tag{3.10b}$$

Example 3: Plant Availability

Suppose an ethylene plant experienced the following outages over a calendar year, find the plant's average overall availability.

Date/time plant out of service	Date/time plant returned to service
2/10/96: 1000 hrs	2/12/96: 1430 hrs
5/28/96: 0800 hrs	6/04/96: 1600 hrs

Solution:

Noting that there were 366 days in 1996 and that the two downtime periods had 52.5 and 176 hours when the plant was out of service, the plant availability in 1996 is calculated as follows:

$$\sum_{i=1}^{n} \mathrm{TTF_i} = 366 \times 24 - (52.5 + 176) = 8555.5 \text{ hr}$$

$$\sum_{i=1}^{n} \mathrm{TTF_i} + \mathrm{TTR_i} = 366 \times 24 = 8784 \text{ hr}$$

$$\bar{A} = \frac{8555.5}{8784} = 0.974$$

Now consider a process consisting of repetitions of the repair-to-failure and failure-to-repair process. Assume that all items are as good as new at time $t = 0$. Similar to the case of nonrepairable items, the reliability parameters for repairable items can be calculated in a systematic manner given that the history of failure times and repair times for each item in the sample throughout the test duration is known.

Example 4:
Calculate the value of MTTF for the 10 components of Figure 3.2.

Solution:
The histories of failure and repair times for the 10 items are summarized in Table 3.5. Note that repair times and failure times are measured from time zero, while TTF represents a time interval.

Calculations are performed in a similar fashion to the previous examples.

Repair Rate $r(t)$:
In its simplest form, the repair rate $r(t)$ may be calculated in a similar fashion to calculating the failure rate. Consider the following notations:

TTR = time-to-repair,

$M(t)$ = number of repairs completed during an interval of time less than or equal to the value of t,

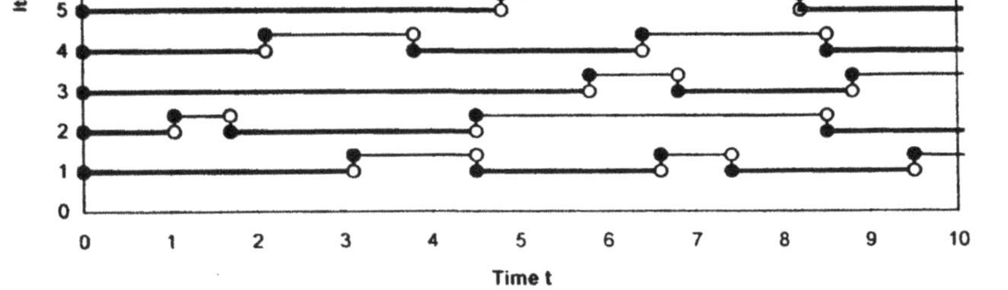

FIGURE 3.2 ***Repair failure sequences. A sequence starting with a solid dot and ending with an open dot represents a start–run–stop sequence for normal ("N") operation. The other, thinner lines ("F") depict items that are in a failed state.***

TABLE 3.5
History of Items States

Component	Repair t	Failure t	TTF
1	0	3.1	3.1
1	4.5	6.6	2.1
1	7.4	9.5	2.1
2	0	1.05	1.05
2	1.7	4.5	2.8
3	0	5.8	5.8
3	6.8	8.8	2
4	0	2.1	2.1
4	3.8	6.4	2.6
5	0	4.8	4.8
6	0	3	3
7	0	1.4	1.4
7	3.5	5.4	1.9
8	0	2.85	2.85
8	3.65	6.7	3.05
9	0	4.1	4.1
9	6.2	8.95	2.75
10	0	7.35	7.35
		Sum of Time Intervals	54.85
		Number of Time Intervals	18
		MTTF	3.05

$G(t)$ = repair probability at time t (i.e., the probability that the repair is completed before time t, given that the item failed at time 0),
$g(t)$ = repair density corresponding to $G(t)$
$r(t)$ = repair rate.
The mathematical representation for $r(t)$ is

$$r(t) = \frac{g(t)}{1 - G(t)} \tag{3.11}$$

where

$$G(t) = \frac{M(t)}{n} \tag{3.12}$$

and

$$g(t) = \frac{G(t + \Delta t) - G(t)}{\Delta t} \tag{3.13}$$

Once again, the calculations contained within the CCPS Database software make a simplifying assumption that all repairs are made in response to an unexpected failure, either while the equipment is on-line or while it is being tested. In reality, repairs may be made as part of a regular schedule, or opportunistically while other items are being repaired when this occurs. It destroys the independence of the data points, making the exponential distribution unsuitable. For this reason, the Database users are provided the option of extracting "nonideal" data.

Aspects of Repair Times:
When dealing with repair times, the approach may not be straightforward. A primary issue is that of definitions: For example, is one interested in the total time the equipment is out of service, or the actual time spent doing the repair? The former is highly dependent on the item's application, while the latter is more inherent to the item itself. These concepts are illustrated in Figure 3.3

For the purposes of the CCPS Database, the following measures are considered:

- Time to restore
- Time to repair—total
- Time to repair—active

"Time to restore" is the interval from which a failure is discovered to the point at which the application is returned to service. This time is mostly dependent on the application itself (location or "TAG#" in CCPS Database nomenclature) of the equipment. The "total repair time" is considered to be the time interval during which a failure is first detected and the item is repaired. In this case, both the item type ("Item #") and "application" (TAG#) are important. The "active repair time," in contrast, strictly depends on the ease of repair for the item itself: for example, whether the pump is a Krupstein Model 34A or a Bentley Model 14C63. In other words, the location of the item in the facility (e.g., depropanizer feed pump) is less relevant, and so the CCPS Database uses the "Item #" for this calculation.

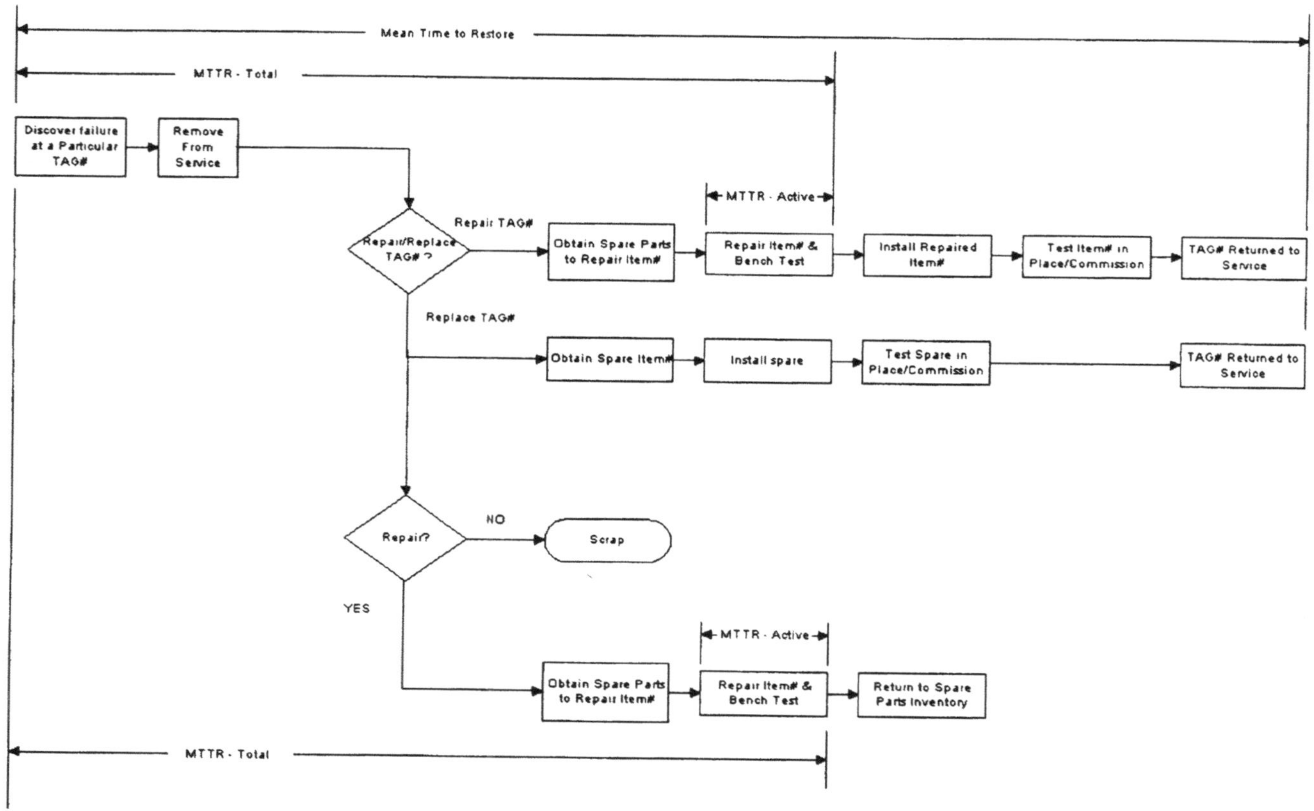

NOTES 1) Mean Time to Restore: Based on tag# Analyst needs to be able to filter on repair or replace method of restoration
2) MTTR - Active. Based on item#.
3) MTTR - Total: Based on item#

FIGURE 3.3. ***Repair times.***

3.4. Cyclical Service

The previous examples have illustrated failure and repair calculations for equipment that is normally in continuous service. For items in cyclical operation, failures *per cycle* and failures *per unit time* may both be meaningful concepts. The calculation of failure per cycle is relatively simple, given a count of the failures, and the number of cycles that have occurred in the surveillance period.

The concept of failure per unit time is more problematic. One must decide whether to include the entire surveillance time or only that time in which the equipment is "active." The former assumes that "degradation" occurs when the equipment is nonactive as well as active. The latter assumes that the equipment is "stressed" only when active. See Section 3.8 for an example of this effect. For the purposes of the CCPS Database, it is assumed that all time, including " nonactive" time, is included in calculations using time.

3.5. Batch Service

The situation encountered in batch service is similar to that in cyclical service, and is calculated in analogous ways. Thus, calculations for failures per batch and failures per unit time are considered valid.

3.6. Standby Service

Failures of equipment in standby service can be calculated in terms of failures per demand (or per start) or as a function of time. A classic example is a relief valve. Systems such as compressors and pumps, however, are generally run in continuous mode, but may also operate in standby. Common examples include backup utility air compressors and firewater pumps. This type of problem was discussed earlier in the censoring analysis.

3.7. Failures Following a Repair

The CCPS Database does not distinguish failures that occur immediately after start or restart from those that occur after the equipment has finished a "break-in" period. Thus, if an analyst wishes to study failures-to-start immediately following maintenance or repair, the failure is most likely dominated by human error, and use of the constant failure rate λ would not be appropriate. Instead, a failure-to-start probability would be preferred. On the other hand, if relief valves were being proof tested at various time intervals, one might question the validity of failure to open on demand values, since time is an important factor. The analyst is urged to take these factors into account when conducting analyses.

3.8. Selecting an Operating Mode

The selection of an appropriate operating mode may not be easy. Consider the following example:

> *At an ammonia truck loading station, flexible hoses are used to unload from a storage tank into trucks. Anywhere from zero to fifteen trucks may be loaded on a given day, each requiring the hose to be hooked and unhooked. The hose is pressure tested monthly and replaced every two years, regardless of apparent condition, earlier if obvious deterioration has taken place.*

Is the hose in continuous or batch, or even cyclic service? The answer is not clear, and the analyst should select the operating mode based on the projected major contributors to failure. For example, if most failures are anticipated to occur because of wear and tear not associated with the process itself (for example, due to exposure to weather, or poor hose storage conditions), then for all practical purposes the hose can be considered as undergoing continuous degradation. If, on the other hand, the failures are mostly related to time of exposure to ammonia, or to damage during coupling and decoupling, then a batch mode should be selected.

3.9. Analysis Based on Statistical Inferences

The aim of reliability analysis is to facilitate the decision making process for reliability engineers. Under ideal conditions, they would be presented with complete information pertaining to failure and repair times for the entire population. A histogram (graphical representation of the data) can then be constructed and the failure or repair distribution determined by fitting the data to some mathematical expression.

When only limited fragmentary data are available, one may resort to statistical inferences to express the reliability parameters of interest in a functional mathematical form. In practice, the process of statistical inference involves two major steps:

1. modeling, in which a probabilistic model is selected to represent the data, and
2. parameter estimation, in which the values of the parameters of the selected model are estimated.

The previous examples have cited the exponential distribution, which is that provided with the CCPS Database software. Other classes of models are described below.

3.9.1. Modeling Reliability Parameters for the Population

It is important to put some effort into deciding the most appropriate probability model for the available data. Once the model is selected, the only freedom left is in estimating the distribution parameters. It should be emphasized, however, that finding the most appropriate distribution is as much a matter of judgment and experience as it is of statistics and mathematics. As mentioned in the introduction to this chapter, experience indicates that some probability distributions are more representative and more effective in certain situations than others; and again, the analyst has a choice of either parametric (e.g., exponential) or nonparametric (e.g., Kaplan-Meier) models. Several interesting distributions and their important characteristics are summarized in Table 3.6. One particular model of interest is the Weibull distribution. It is by far the most widely employed model and therefore is discussed in more detail in the following section.

3.9.2. The Weibull Distribution

The use of this distribution to model the time-to-failure variable originated in fatigue studies and is of particular practical interest in reliability analysis because of its characteristics. The distribution is flexible enough to accommodate increasing, constant, or decreasing hazard. It is also mathematically appealing because of its simplicity, in addition to being amenable to graphical presentation.

The two-parameter form of the Weibull distribution is described as

$$F(t) - 1 - \exp\left[-\left(\frac{t}{\eta}\right)^{\beta}\right] \tag{3.14}$$

where η (characteristic life) is the scale parameter described as the value of t at which there is an approximately two-thirds probability that the component will have failed; η can be considered an indirect measure of the overall reliability; β is a shape parameter related to the behavior of the hazard function, that is,

$\beta = 1$: the failure rate $h(t)$ is constant and the distribution becomes the exponential distribution,
$\beta > 1$: $h(t)$ is increasing (normally is less than 4),
$\beta < 1$: $h(t)$ is decreasing (normally is greater than 0.5).

At $\beta = 3.44$ the distribution becomes a close approximation of the normal distribution.

By estimating the value of the parameters η and β, we determine which particular member of the possible family of curves of this distribution represents the data under consideration. Parameter estimation for the Weibull distribution can be performed graphically or analytically, based on the maximum likelihood technique. Special computer programs are available for the latter purpose. Descriptions of such programs are left to the references.

3.9.3. Graphical Method for Estimating the Weibull Parameters

Parameter estimation here describes the process of obtaining estimates for η and β from observed failure data. The graphical solution should address the nature of the data, whether it is complete or censored. In either case, the principle is the same.

The proportion failed in the observed sample at a particular time t is used as an estimate of the function $F(t)$ at that time. A graph can be constructed by plotting these estimates at each observed failure time. The objective is to find the parameters of the Weibull distribution that *best fit* the plotted points. Special graph paper, in which the axes are transformed in a way that the plotted points become linear, should be used to overcome the difficult task of ensuring that the plot is a Weibull curve. A basic concept underlines the way the data are analyzed. This concept considers the data of a life test as consisting of a number (n) of similar components that were tested until k of them had failed ($k \leq n$). The concept is adopted even though, in practice, the results of the test are not obtained that way. The Weibull process is illustrated in Chapter 4.

The analyst should be forewarned that the uncertainty on the estimate of the Weibull parameters can be quite large, especially for small samples. This makes it difficult to use inferences for decisions without some knowl-

TABLE 3.6
Common Probability Distributions

Distribution	Failure Density Function $f(t)$	Parameter Identification	Applicability in Reliability Analysis
Exponential	$\lambda \exp(-\lambda / t)$	λ = constant failure rate = $f(t)/R(t)$	• Applicable to data in the absence of other information • Describes a constant hazard rate
Normal	$\frac{1}{\sigma\sqrt{2\pi}} \exp\left[-\frac{(t-\mu)^2}{2\sigma^2}\right]$	μ = mean σ = standard deviation	• Applicable to wearout failures and repair times • Restricted to increasing hazard rate
Lognormal	$\frac{1}{\sigma t\sqrt{2\pi}} \exp\left[-\frac{(\ln t-\mu^*)^2}{2\sigma^2}\right]$	μ^* = mean σ= standard deviation	• To be used when deviations from the model value are by factors, proportions, or percentages rather than absolute values • Applicable to fatigue failures, repair times, and hazard rates • Restricted to increasing hazard rate
Weibull (2-parameter)	$\frac{\beta}{\eta}\left(\frac{t}{\eta}\right)^{\beta-1} \exp\left[-\left(\frac{t}{\eta}\right)^{\beta}\right]$	η= characteristic life β = shape factor	• Applicable to any failure data since it can handle decreasing, constant and increasing rates
Rectangular (uniform)	$\frac{1}{b}$ mean $= a + \frac{b}{2}$	a = location parameter b = scaling factor	• Used to give random variables a uniform distribution over a specified interval for simulation purposes
Gamma	$\frac{1}{b\Gamma(a)}\left(\frac{t}{b}\right)^{a-1} \exp\left(-\frac{t}{b}\right)$	a = shape factor b = scaling factor	• Can handle decreasing, constant or increasing data, but less used than the Weibull distribution
Pareto	$at^{-(a+1)}$	a = shape factor	• Usually used in discrete form to describe the distribution of number of failures
Extreme Value	$\frac{1}{b}\exp\left(\frac{t-a}{b}\right)$ $\times \exp\left[-\exp\left(\frac{t-a}{b}\right)\right]$	a= location parameter b= scaling factor	• Has very limited use to investigate extreme and rare values of the phenomenon

edge of the qualitative descriptions of the failures; otherwise gross errors may result.

3.9.4. The Exponential Distribution

An exponential distribution is one in which failures are assumed to occur randomly over time. This distribution is equivalent to a Weibull distribution with β equal to 1. There are several reasons why failures may take an exponential (random) form, and why this is the default distribution assumed by the CCPS Database software, including

- Failures due to natural events such as lightning
- Random human errors

An exponential distribution may be suggested for other, nonrandom reasons, however. If there are three or more prominent failure mechanisms for an equipment item, the "blended" Weibull data will also tend to produce a combined β near 1. In such a case, the data are more properly segregated into their separate component Weibull curves in order to obtain more meaningful results. The exponential assumption may also be adopted to reflect the effect of time-based maintenance systems in making failures *appear* to be exponentially distributed. Lastly, the exponential approach may simply be the most mathematically convenient.

Similarly, in the past, the exponential distribution was utilized to describe the entire life-cycle of equipment. A typical life-cycle may include an "infant mortality" stage ($\beta < 1$), a "useful life" stage (characterized by random failures; $\beta = 1$), and a "wear out" stage ($\beta > 1$). These data may also combine to form an apparent exponential distribution, thus misleading the analyst. The proper handling of such data is illustrated in Section 4.10.

3.9.5. Confidence Limits for Reliability Parameters

When we estimate the characteristics of a population from a given set of sample data, it is difficult to be certain about how well the sample represents the population. Generally, as the sample size increases, the values of the sample and those of the population become closer. Although we cannot be certain that a sample is representative of a population, it is usually possible to associate a degree of assurance with a sample characteristic. That degree of assurance is called *confidence*. Confidence is defined as the level of certainty associated with conclusions based on the results of sampling.

When we associate a confidence limit with a probabilistic parameter (e.g., reliability), it indicates that we are $(1-\alpha)$ confident, or $100(1-\alpha)\%$ confident, that the true population reliability lies between a lower and an upper possible values. The *level of significance*, α, must lie between zero and one, and $(1-\alpha)$ is a limit that refers to one end of the reliability. The range between α and $(1-\alpha)$ is called *the confidence interval*. Confidence limits could be one sided (confidence $= 1-\alpha$), or double sided (confidence $= 1-2\alpha$).

In reliability analysis, there are two approaches to determine confidence limits. We shall provide a general overview of those approaches; the details are left to the references.

Classical Confidence Limits

The analysis here is based on having N random samples from the population. Let θ denote the characteristic of the population in which we are interested. The same characteristic can be calculated from data of each sample. The calculated values will all be different and none of them represents the true value of θ. Thus, the calculated values will have a certain probability density function out of which we are trying to determine the best estimate (S) for the true value of θ with a certain degree of confidence. In other words, the classical approach uses the sampling distribution to determine two numbers, $s_1(\theta)$ and $s_2(\theta)$, as a function of θ, such that

$$\Pr(S \le s_2(\theta)) = 1-\alpha \tag{3.15}$$

and

$$\Pr(S \ge s_1(\theta)) = 1-\alpha \tag{3.16}$$

therefore,

$$\Pr[s_1(\theta) \le S \le s_2(\theta)] = 1-2\alpha \tag{3.17}$$

Generally, $s_1(\theta)$ and $s_2(\theta)$ are often referred to as Θ_1 andΘ_2, respectively.

Consider the case where one is dealing with life tests of N identical items and the life test is terminated at the time of the kth failure ($k \le N$). When our reliability characteristic of interest for the population (θ) is the mean time to failure (MTTF), the procedure for determining the upper and lower confidence limits may be as follows.

Assume N items with an exponential TTF distribution. The estimated value of (θ) from observed data is

$$S = \frac{(N-k)t_k + \sum_{i=1}^{k} t_i}{k} \tag{3.18}$$

The mathematical expression $(2kS/\theta)$ follows a chi-square distribution with $2k$ degrees of freedom. Based on that information, and some mathematical derivations, the lower and upper confidence limits, Θ_1 and Θ_2 are given by

$$\Theta_1 = \frac{2kS}{\chi^2_{1-\alpha}(2k)} \tag{3.19}$$

$$\Theta_2 = \frac{2kS}{\chi^2_{\alpha}(2k)} \tag{3.20}$$

The values of Θ_1 and Θ_2 give $100(1-\alpha)\%$ as the lower and upper confidence limits. The range (Θ_2, Θ_1) is the $100(1-2\alpha)$ confidence interval.

Example 5

Assume 30 identical items having the above censoring with $k = 20$. The 20th failure has a TTF of 39.89 minutes ($t = 39.89$), and the other 19 TTFs are

0.26; 1.49; 3.65; 4.25; 5.43; 6.97; 8.09; 9.47; 10.18; 10.29;
11.04; 12.07; 13.61; 15.07; 19.28; 24.04; 26.16; 31.15; 38.70;

Find the 95% two-sided confidence limits for the MTTF.

Solution:

$$N = 30;\ k = 20;\ t = 39.89;\ \alpha = 0.025$$

Because of the lack of information about TTF of the last 10 items, we assign a TTF value to each item equal to that of the 19th item. Therefore,

$$S = [(30-20) \times 39.89 + 291.09]\,/\,20 = 34.5 \text{ min.}$$

From the chi-square tables:

$$\chi^2_{0.025}(2k) = \chi^2_{0.025}(40) = 59.3417$$

$$\chi^2_{1-\alpha}(2k) = \chi^2_{1-\alpha}(40) = 24.4331$$

Equations (3.19) and (3.20) give

$$\Theta_1 = 2 \times 20 \times 34.5/59.3417 = 23.26$$

$$\Theta_2 = 2 \times 20 \times 34.5/24.4331 = 56.48$$

Then, the confidence interval is $23.26 \le \theta \le 56.48$. In other words, we are 95% confident that the MTTF (θ) is in the interval 23.26 to 56.48 minutes.

Bayesian Approach

The approach of classical statistics for determining reliability parameters has one shortcoming. It does not take into account past experience or information. This is particularly important in the case of analyzing an improved item. One method of combining *a priori* experience with posterior test data is the Bayesian approach. It is based on Bayes's theorem which states:

Posterior Probability is proportional to prior probability × Likelihood.

One of the important applications of the Bayesian approach is in binomial testing situations, in which the possible outcomes of the tests are either success or failure. Similar to the classical approach, some mathematical expressions are derived to determine confidence limits for the reliability characteristics of interest. Details of those procedures are left to the references.

The intent is to operate the CCPS Database software such that automated failure analysis will be allowed only when some critical amount of data has been collected. The Database may require some multiple of the expected time to a single failure. For example, if the expected failure rate for an equipment item is once in 10 years, then 10×, or 100 years, of equipment history may be specified. Nonetheless, methods do exist for analyzing smaller samples of data, as described above, and there are even approaches for assessing failure rates when no failures have occurred (see References).

References

ARINC Research Corp., "Reliability Engineering," W.H. Von Alven, Ed., Prentice-Hall, Englewood Cliffs, NJ (1964).

Ascher, H. and Feingold, H., "Repairable Systems Reliability," Marcel Dekker, New York (1984).

Collett, D., "Modelling Survival Data in Medical Research," Chapman and Hall, London (1994).

Doty, L. A., "Reliability for the Technologies," 2nd ed., Industrial Press, New York (1989).

Henley, E. J., "Reliability Engineering and Risk Assessment," Prentice-Hall, Englewood Cliffs, NJ (1981).

Nelson, W. B., "Applied Life Data Analysis," John Wiley & Sons, New York (1982).

Nelson, W. B., "Volume 6: How to Analyze Reliability Data," American Society for Quality Control, Milwaukee (1983).

4 Example Applications

4.1 Introduction

The CCPS Database will allow users to extract raw performance data, selected by the parameters of interest to the user. Simple calculations will also be provided within the scope of the Database (e.g. average failure rate, mean times to repair). Some more sophisticated analyses will be beyond the scope of the software, in large part because the type of technique to be employed is highly dependent on the nature of the system being analyzed.

In Chapter 3, many data analysis methods were introduced, including several types of failure probability distributions. The failure of any particular part of a system to be studied may be reasonably characterized by one, or possibly more, of these failure distributions. Given the large number of parts in a system, the failure distribution for the system as a whole may or may not be dominated by the distribution of one of the parts in the system. The Database software assumes exponential failure distributions for calculation purposes. In many cases, however, the type of failure distribution that characterizes a complex system is not known a priori. Therefore, the exercise of determining the precise nature of the failure distribution is largely left to the Database user, if something other than the exponential distribution is suspected. The reader is invited to explore the references for more information on selecting an appropriate failure distribution. Section 4.8 also provides an example of sorting through combinations of failure distributions. In any case, it is appropriate to provide guidance and examples for those users who are interested in more than the "standard" sorts of calculations.

This chapter consists of two main themes:

- Manipulation of raw data to determine failure distributions and various measures of reliability (Sections 4.2–4.8)
- Application of reliability data to plant situations (Sections 4.9–4.13)

4.2. Conducting a Reliability Analysis—Pump Example

As noted in Chapter 3, the Weibull distribution function in particular has found wide favor in industry due to its flexibility over a wide variety of behaviors, and the fact that it is amenable to graphical analysis. Following is an example of how raw data can be interpreted using a Weibull approach, then further manipulated to allow decision making at the plant level.

Definitions

It is useful at this point to revisit some earlier terms:

t = Time, measured from the beginning (normally zero) to the end of the surveillance period. It is also used to denote the time interval between tests, in availability calculations

$R(t)$ = Reliability, it is the probability that an item is functioning at a given point in time

$f(t)$ = Failure density function. The failure rate will also be referred to as "λ" when there is no known dependence of the failure rate on the time in service

β = Weibull "shape parameter"

η = Weibull "characteristic life"

Raw Data Analysis—Weibull

A plant is in the process of specifying equipment for a unit. The specification for one of the pumps is considered a key, since the pump is to be in a severe service and the process shuts down upon loss of flow from the pump. The database has been searched for histories of comparable pumps in similar service, and records for 10 such pumps have been obtained:

Pump 1: Has operated for 12 months with no failures
Pump 2: Failed after 21 months of operation
Pump 3: Has operated for 12 months with no failures
Pump 4: Has operated for 14 months with no failures
Pump 5: Has operated for 36 months with no failures
Pump 6: Failed after 15 months of operation
Pump 7: Failed after 6 months of operation
Pump 8: Has operated for 48 months with no failure
Pump 9: Has operated for 24 months with no failure
Pump 10: Failed after 13 months of operation

There are many approaches to process data into cumulative failure form; these are described in the references listed at the end of this chapter.

For this example the "median rank" approach is used (see Section 4.3), which is arguably the most popular method in use. The raw performance data above are then rearranged as follows:

Step 1. Write down the *k* failure ages (time-to-failure) of the items in an ascending order of magnitude, where t_1 is the smallest value and t_k is the largest value.

Step 2. The "Auth" formula can be used to develop "adjusted ranks" to account for censoring of data:

Adjusted Rank = [(Rank)(Previous Adjusted Rank) + (n + 1)]/[(Rank) + 1],

where n becomes the sum of failure and nonfailure data points.

Step 3. "Benard's approximation" can then be used to calculate a median rank for each data point, as follows: $F(t_i) = (i - 0.3)/(n + 0.4)$, where i becomes the Adjusted Rank. Also, many users prefer to describe the distribution function in terms of percentages, rather than fractions, of equipment failed. This approach is used in Table 4.1, which simply uses values of $F^\wedge(t_i) = 100F(t_i)$.

TABLE 4.1
Pump Failure Times

Item No.	Rank	t (months)	t (years)	Adjusted Rank	$F^\wedge(t_i)$
Pump 7	10	6*	0.50*	1	6.7
Pump 3	9	12	1.00		
Pump 1	8	12	1.00		
Pump 10	7	13*	1.08*	2.25	18.7
Pump 4	6	14	1.17		
Pump 6	5	15*	1.25*	3.71	32.8
Pump 2	4	21*	1.75*	5.17	46.8
Pump 9	3	24	2.00		
Pump 5	2	36	3.00		
Pump 8	1	48	4.00		

* Indicates that a failure took place at this time

Step 4. Plot the data on Weibull graph paper and draw the best straight line through the plotted points (see Figure 4.1).

Step 5. Draw a line parallel to the one drawn in Step 3, in such a way that it goes through the "Origin" point located at the top center of the Weibull graph paper. The point of intersection of this parallel line with the β estimation scale on the left side of the paper gives the estimate of the shape parameter.

Step 6. To estimate the characteristic life η, drop a vertical line down from the original plot at the point where the plot is at a value of $F^\wedge(t_i) \sim 63\%$ on the Y axis. The value at the X axis is the parameter η. At this stage, the functional form of the Weibull model that best represent the data is determined.

Plotting the data above results in a shape parameter β of 1.8 and a characteristic life η of 2.2 years. With this information, many judgments can be made. Some possibilities are provided in the following sections. The user should note that the Weibull parameters just derived are based on relatively few data points, and are likely to have a high degree of variability. It is possible to calculate confidence intervals for Weibull parameters, but that is not part of this current exercise.

There are also various types of Weibull plot configurations, which frequently use axes that are the reverse of the one used in this example (and having inverted instructions from those above). Each type has its own advantages.

A Warning About the Use of Weibull Distributions

Figure 4.1 illustrates a simple Weibull case. Before venturing too far into this area, the reader should consult other references, particularly Reference 1 at the end of this chapter. There are a number of complications that can distort the results. For example, if the data on a Weibull plot appear to form a curve rather than a straight line, one or more of the following is likely:

- ***Failures can not occur until a period of time has elapsed. If this is the case, the Weibull procedure above should be modified so that the time axis values are (actual time – minimum failure time) rather than (actual time).***
- ***There is more than one significant failure mechanism, and the plotted curve represents a mix of two or more individual Weibull lines (see Figure 4.5 as an example). From previous experience, it is known that different failure causes, and therefore different characteristic Weibull plots, can result from the following:***

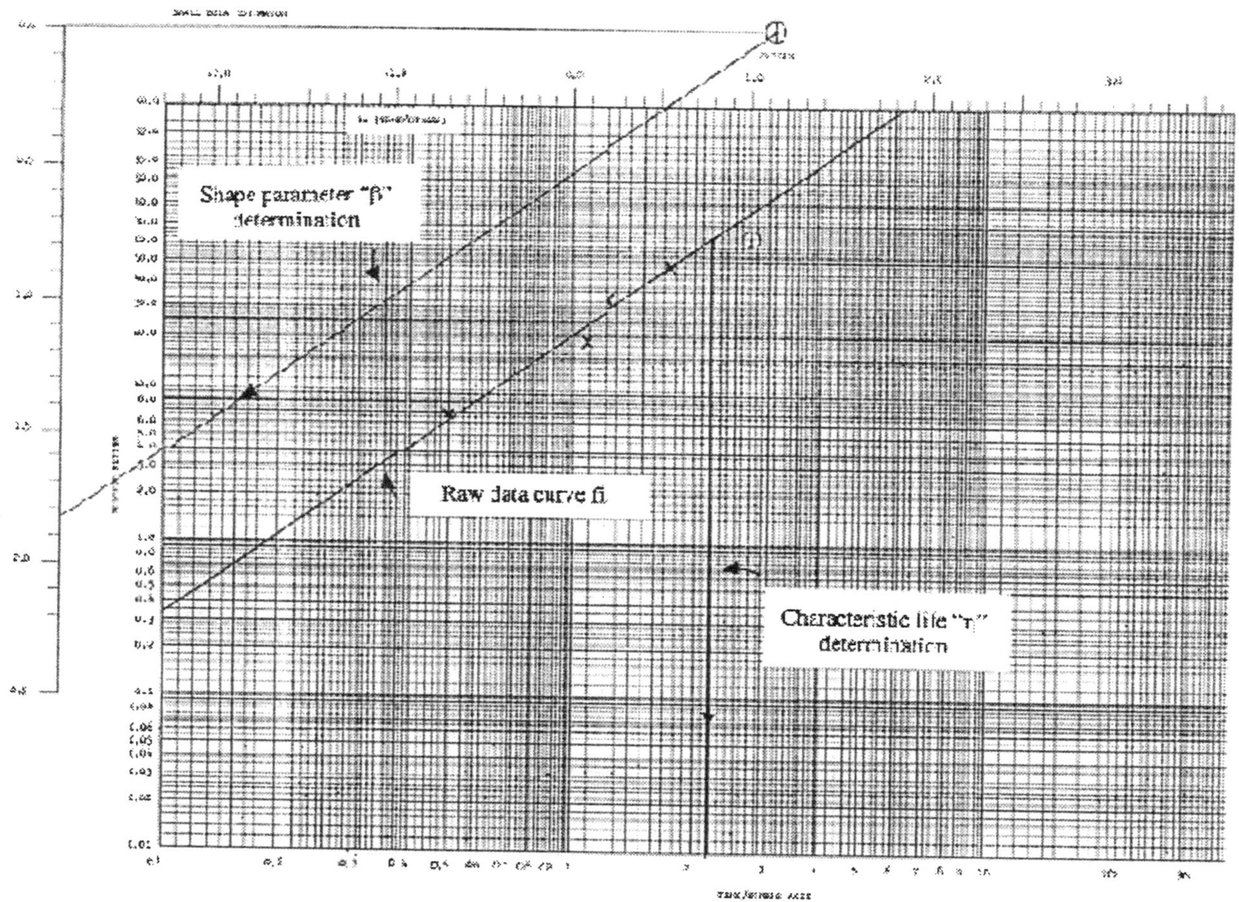

FIGURE 4.1. ***Weibull pump data plot.***

- *data being collected at different times in the lives of the equipment items*
- *different manufacturers of the same item*
- *different types/sizes of the same item within the same manufacturer's line*

Hence, mixing too many disparate data sources together may tend to cloud the results—this is typically observed as a β *approaching 1.*

4.3. Right-Censoring

A simple numerical example is presented below for the purpose of demonstrating how data censoring affects the analysis of reliability parameters. It is useful at this point to define some terms:

$F(t)$ = Failure distribution: the probability that an item experiences the first failure during time interval 0 to t, given that it is normal at time 0.
i = A particular data sample point
N = Number of random samples from a population
$R(t)$ = Reliability distribution; equal to $1 - F(t)$
t = Time, measured from zero to the end of the surveillance period

Consider the lifetime to failure for bearings given in Table 4.2. The original number of bearings exposed to failure is 202; however, between each failure some of the bearings are taken out of service before failure has occurred. Calculate $F(t)$ and $R(t)$.

Sample Calculations:
At lifetime to failure = 716
Value in Column (4) = 1 × (202 – 4.47)/156 = 1.27
Cumulative number of failures expected = 4.47 + 1.27 = 5.74
$F(t)$ = 5.74/202 = 0.0284
$R(t)$ = 1 – 0.0284 = 0.9716 [based on Eq. (3.3)].

Note: The general approximation $F(t_k) = k/N$, is applicable when all items concerned continue to failure. However, when items are taken out while being in a normal state, a correction should be made to account for data censoring. The necessary adjustment is expressed by

$F(t)$ = cumulative number of failures expected if all original items have been allowed to proceed to failure divided by the number of items exposed to failure, at time t.

TABLE 4.2
Bearing Test Data and Results

(1) Lifetime to Failure (hr)	(2) No. of Failures	(3) No. Exposed to Failure	(4)* No. of Failures Expected if Orig. Population had been Allowed to Proceed to Failure	(5) Cumulative No. of Failures Expected	(6) $F(t)$	(7) $R(t)$
141	1	202		1.0000	0.0050	0.9950
337	1	177	1.1356	2.1356	0.0106	0.9894
364	1	176	1.1356	3.2712	0.0162	0.9838
542	1	165	1.2044	4.4756	0.0222	0.9778
716	1	156	1.2662	5.7418	0.0284	0.9716
765	1	153	1.2827	7.0245	0.0348	0.9652
940	1	144	1.3540	8.3785	0.0415	0.9585
986	1	143	1.3540	9.7325	0.0482	0.9518

*Value in column (4) = that in column (2) × [202 – value in preceding row of column (5)] divided by value in column (3). Information about this computational procedure is provided in the note on page 52.

At t_1 the proportion expected to fail is k_1/N_1. The number that would have failed at t_2, provided the original number (N_1) had been allowed to proceed to failure is

$$(k_2/N_2)(N_1 - k_1).$$

MEAN, MEDIAN, AND MODE

The estimated value of $F(t)$ for the ith failure is usually one of the measures of central tendencies of the Beta distribution at that particular point. The measures of central tendencies include the mean, the median, and the mode. For the ith failure these are given by

$$\text{Mean rank} = i/(N + 1),$$

$$\text{Median rank} = (i - 0.3)/(N + 0.4), \text{ and}$$

$$\text{Mode} = (i - 1)/(N - 1).$$

To illustrate the effect of the type of tendency selected, consider some of the data from Table 4.3 and the formulas above.

TABLE 4.3
Comparison of Tendencies on Table 3.2 Data

(1) Lifetime to Failure (hr)	(2) No. Exposed to Failure	(3) No. of Failures Expected if Orig. Population had been Allowed to Proceed to Failure	(4) Cumulative No. of Failures Expected	(5) $F(t)$ [Mean]	$F(t)$ [Median]	$F(t)$ [Mode]
141	202		1.0000	0.0050	0.0035	0.0000
337	177	1.1356	2.1356	0.0106	0.0091	0.0056
364	176	1.1356	3.2712	0.0162	0.0147	0.0113
542	165	1.2044	4.4756	0.0222	0.0206	0.0173
716	156	1.2662	5.7418	0.0284	0.0269	0.0236
765	153	1.2827	7.0245	0.0348	0.0332	0.0300
940	144	1.3540	8.3785	0.0415	0.0399	0.0367
986	143	1.3540	9.7325	0.0482	0.0466	0.0434

Which measure to use is a matter of judgment, previous practice, and circumstances. In general, the median approach is preferred for non-symmetrical distributions, since most equipment life data are not equally distributed on each side of the mean.

4.4. MTTF by Numerical Integration

Table 4.4 shows failure data for 250 transistors, all of which failed during the test period. Calculate the unreliability $F(t)$, the failure rate $h(t)$, the failure density $f(t)$, and the MTTF using Eqs. (3.5)–(3.7).

TABLE 4.4
Failure Data for Transistors

Time to Failure t (min)	Cumulative Failures k	Time to Failure t (min)	Cumulative Failures k
0	0	230	143
20	9	400	160
40	23	900	220
60	50	1200	235
90	83	2500	240
160	113	>2500	250

TABLE 4.5
Transistor Reliability, Unreliability, Failure Density, and Failure Rate

t	$L(t)$	$R(t)$	$F(t)$	$k(t + \Delta t) - k(t)$	Δt	$f(t)$	$h(t)$
0	250	1	0	9	20	0.0018	0.0018
20	241	0.9640	0.0360	14	20	0.002800	0.002905
40	227	0.9080	0.0920	27	20	0.005400	0.005947
60	200	0.8000	0.2000	33	30	0.004400	0.005500
90	167	0.6680	0.3320	30	70	0.001714	0.002566
160	137	0.5480	0.4520	30	70	0.001714	0.003128
230	107	0.4280	0.5720	17	170	0.000400	0.000935
400	90	0.3600	0.6400	60	500	0.000480	0.001333
900	30	0.1200	0.8800	15	300	0.000200	0.001667
1200	15	0.0600	0.9400	5	1300	0.000015	0.000256
2500	10	0.0400	0.9600	0	−2500		

Sample Calculations:

n =250 transistors.

At time t = 20 minutes,

$L(20) = n - k(20) = 250 - 9 = 241$ transistors

$R(20) = L(20)/n = 241/250 = 0.964$

$F(20) = 1 - R(20) = 1 - 0.964 = 0.036$

$k(40) - k(20) = 23 - 9 = 14$ transistors

$\Delta t(20) = 40 - 20 = 20$ minutes

$f(20) = 14(1/250 \times 20) = 0.0028$

$h(20) = 0.002/0.964 = 0.0029$

Middle point in time interval $t(0, 20) = 10$

From a numerical integration of Eq. (3.8):

$$\text{MTTF} = (10 \times 0.0018 \times 20) + (30 \times 0.0029 \times 20) + \cdots$$
$$+ (1850 \times 0.0002 \times 1300) = 501.4$$

4.5. Reliability Calculations for Repaired Items

Reliability can be considered the probability that an item has survived a certain time without failure. If an item is repairable, it may experience a prior failure but still be available at a given time. Under these conditions, however, it would normally not be considered reliable.

Consider a process consisting of repetitions of the repair-to-failure and failure-to-repair process. Assume that all items are as good as new at time $t = 0$. Similar to the case of nonrepairable items, the reliability parameters for repairable items can be calculated in a systematic manner given that the history of failure times and repair times for each item in the sample throughout the test duration is known.

The reliability $R(t)$ and failure rate $h(t)$ can be calculated from Eqs. (3.5) and (3.6), respectively, following the same procedure described for nonrepairable items. It is important to notice, however, as was previously mentioned, that the value of n in this case is greater than the sample size, since the same item in the sample may be counted more than once if it experiences more than one failure during the test period. For the same reason, $L(t)$ can be calculated by determining the number of items with TTF $\geq t$ at any time t. The following numerical example illustrates the mathematical procedure.

Example
Calculate values for $R(t)$, $F(t)$, $f(t)$, and $h(t)$ for the 10 components of Figure 4.2.

Solution
From Fig. 4.2, the histories of failure and repair times for the 10 items are summarized in Table 4.6. Note that repair times and failure times are measured from time zero, while TTF represents a time interval.

Note that the value of n here is equal to the number of rows, 18, in Table 4.6. Calculations are performed in a similar fashion to the previous examples.

4.6. Calculation of MTTR by Numerical Integration

The following repair times (TTRs) for the repair of electric motors have been recorded. Use this data to calculate the values for $G(t)$, $g(t)$, $r(t)$, and MTTR.

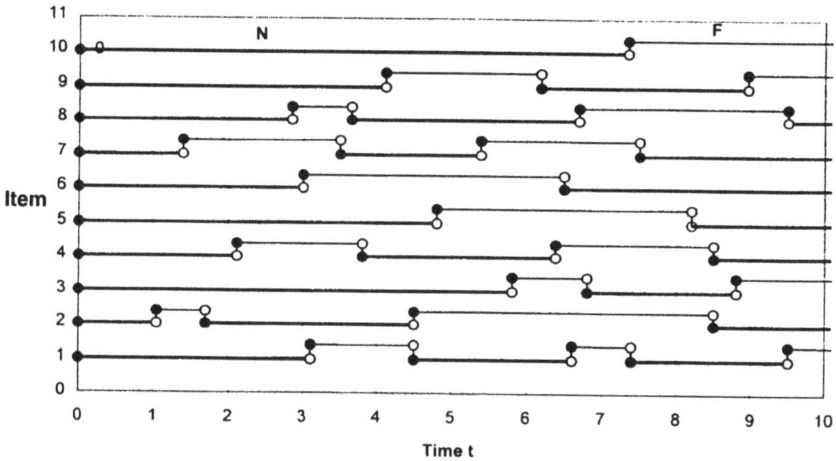

FIGURE 4.2. ***Repair failure sequences (N = Normal, F = Failure).***

TABLE 4.6
History of Items States

Component	Repair *t*	Failure *t*	TTF
1	0	3.1	3.1
1	4.5	6.6	2.1
1	7.4	9.5	2.1
2	0	1.05	1.05
2	1.7	4.5	2.8
3	0	5.8	5.8
3	6.8	8.8	2
4	0	2.1	2.1
4	3.8	6.4	2.6
5	0	4.8	4.8
6	0	3	3
7	0	1.4	1.4
7	3.5	5.4	1.9
8	0	2.85	2.85
8	3.65	6.7	3.05
9	0	4.1	4.1
9	6.2	8.95	2.75
10	0	7.35	7.35

TABLE 4.7
Results of the Reliability Analysis

t	$L(t)$	$R(t)$	$F(t)$	$k(t + \Delta t) - k(t)$*	Δt	$f(t)$	$r(t)$
0	18	1.0000	0.0000	0	1	0.000000	0.000000
1	18	1.0000	0.0000	3	1	0.166667	0.166667
2	15	0.8333	0.1667	8	1	0.444444	0.533333
3	7	0.3889	0.6111	3	1	0.166667	0.428571
4	4	0.2222	0.7778	2	1	0.111111	0.500000
5	2	0.1111	0.8889	1	1	0.055556	0.500000
6	1	0.0556	0.9444	0	1	0.000000	0.000000
7	1	0.0556	0.9444	1	1	0.055556	1.000000
8	0	0.0000	1.0000	0	1	0.000000	
9	0	0.0000	1.0000	0			

* $k(t + \Delta t) - k(t)$ is calculated as: $L(t) - L(t + \Delta t)$.

TABLE 4.8
Electric Motors Repair Times

Repair No.	Time (hr)	Repair No.	Time (hr)
1	3.3	10	0.8
2	1.4	11	0.7
3	0.8	12	0.6
4	0.9	13	1.8
5	0.8	14	1.3
6	1.6	15	0.8
7	0.7	16	4.2
8	1.2	17	1.1
9	1.1		

TABLE 4.9
Results of Electric Motors Repair Time Analysis

t (TTR)	No. of Completed Repairs M(t)	*G*(*t*)	*g*(*t*)	*r*(*t*)
0.0	0	0.0000	0.0000	0.0000
0.5	0	0.0000	0.9412	0.9412
1.0	8	0.4706	0.5882	1.1111
1.5	13	0.7647	0.2353	1.0000
2.0	15	0.8824	0.0000	0.0000
2.5	15	0.8824	0.0000	0.0000
3.0	15	0.8824	0.1176	1.0000
3.5	16	0.9412	0.0000	0.0000
4.0	16	0.9412	0.1176	2.0000
4.5	17	1.0000		

Sample Calculations
The repair times in Table 4.8 range in value approximately between zero and 4.5 hours. By choosing a constant increment of time equal to 0.5 hours, repair times are arranged in Table 4.9 in an ascending order of magnitude. The number of completed repairs $M(t)$ associated with any value of TTR is determined by calculating the frequency of times in Table 4.8 which are less than or equal to that value.

For TTR = 1.0 hour

$M(t) = 8$

The total number of repairs $n = 17$

$G(t) = 8/17 = 0.4706$

$\Delta = 0.5$ hours

$g(t) = (0.7647 - 0.4706) \times (1/0.5) = 0.5882$

$r(t) = 0.5882/(1 - 0.4706) = 1.11$

$$\text{MTTR} = (0.25 \times 0.0) + (0.75 \times 0.9412) + \cdots + (4.25 \times 0.1176) = 1.3676 \text{ hours}$$

The average repair time, MTTR, can also be calculated from

$$\text{MTTR} = (1/17) \times (3.3 + 1.4 + 0.8 + \cdots + 1.1) = 1.3588 \text{ hours.}$$

4.7. Fitting a Weibull Distribution

If the Weibull distribution is fitted to the data from Table 3.2, then the cumulative distribution function, $F(t)$, takes the form

$$F(t) = 1 - \exp\left[-\left(\frac{t}{\eta}\right)^{\beta}\right]$$

After taking logs on both sides, the equation can be reduced to

$$\ln[-\ln(1 - F)] = \beta \ln t - \beta \ln \eta$$

where the parameters β and η can be found by the least squares method.

Let $Y = \ln[-\ln(1 - F)]$
$X = \ln(t)$
$b_0 = \beta \ln \eta$
$b_1 = \beta$

The equation to be regressed is $Y = b_0 + b_1X$, where b_0 and b_1 can be found by simple linear regression.

The quality of the fit is determined by traditional statistics like the coefficient of determination R^2, the mean-squared error, the t-statistics for the parameters, tests for residuals, etc.

If the hypothesis that the shape parameter of the Weibull distribution is $\beta = 1$, then the time between failures follows an exponential distribution.

SAS has a procedure called RELIABILITY which facilitates the computation of these statistics.

Weibull Parameter Estimates				
			Asymptotic Normal 95% Confidence Limits	
Parameter	Estimate	Standard Error	Lower	Upper
Scale	585.3553	80.6620	446.8111	766.8585
Shape	1.0797	0.1217	0.8657	1.3465

The 95% confidence interval for the *shape* parameter contains the value of 1 and therefore the exponential distribution is acceptable.

The mean of the distribution (MTTF) is 568.39. The MTTF calculated from the exponential distribution was 568.82, which shows that even though the exponential distribution is a very good approximation, there is still an effect due to the fact that β is not exactly one (see Figure 4.3).

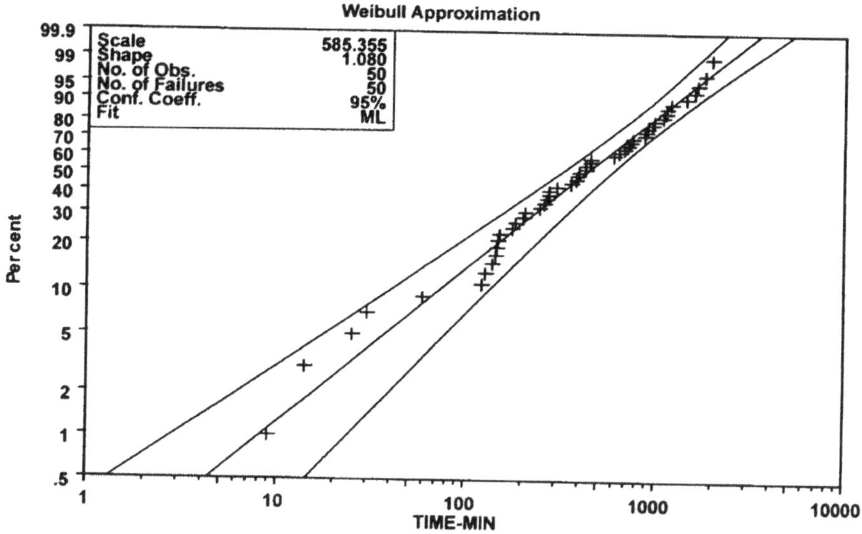

FIGURE 4.3. ***Weibull example.***

The reader is reminded that Weibull analysis was born as a graphical analysis. The latest version of BS 5760 Part 2 reminds the analyst of this and recommends the use of graph paper as an alternative to analytical solutions alone. This enables the analyst to retain a "feel" for the situation, spotting outliers easily and circumventing the uncertainty problems that are associated with small sample populations.

4.8. Combinations of Failure Distributions

In many cases, equipment failure rates do not follow the consistent Weibull failure rate distribution for the reasons described in Section 4.1. In such cases the Weibull plot is still very useful, and can give clues as to the mechanisms behind the equipment failure.

Suppose that failure data are available for 2000 pressure vessels, and that the failure of interest is a leak to atmosphere. Consider the Weibull plot of these data in Figure 4.4 (note that "representative" data points are plotted for times over 5 years):

If the data followed a fixed Weibull distribution, this plot would be a straight line. As can be seen, however, this is not the case. What is the explanation?

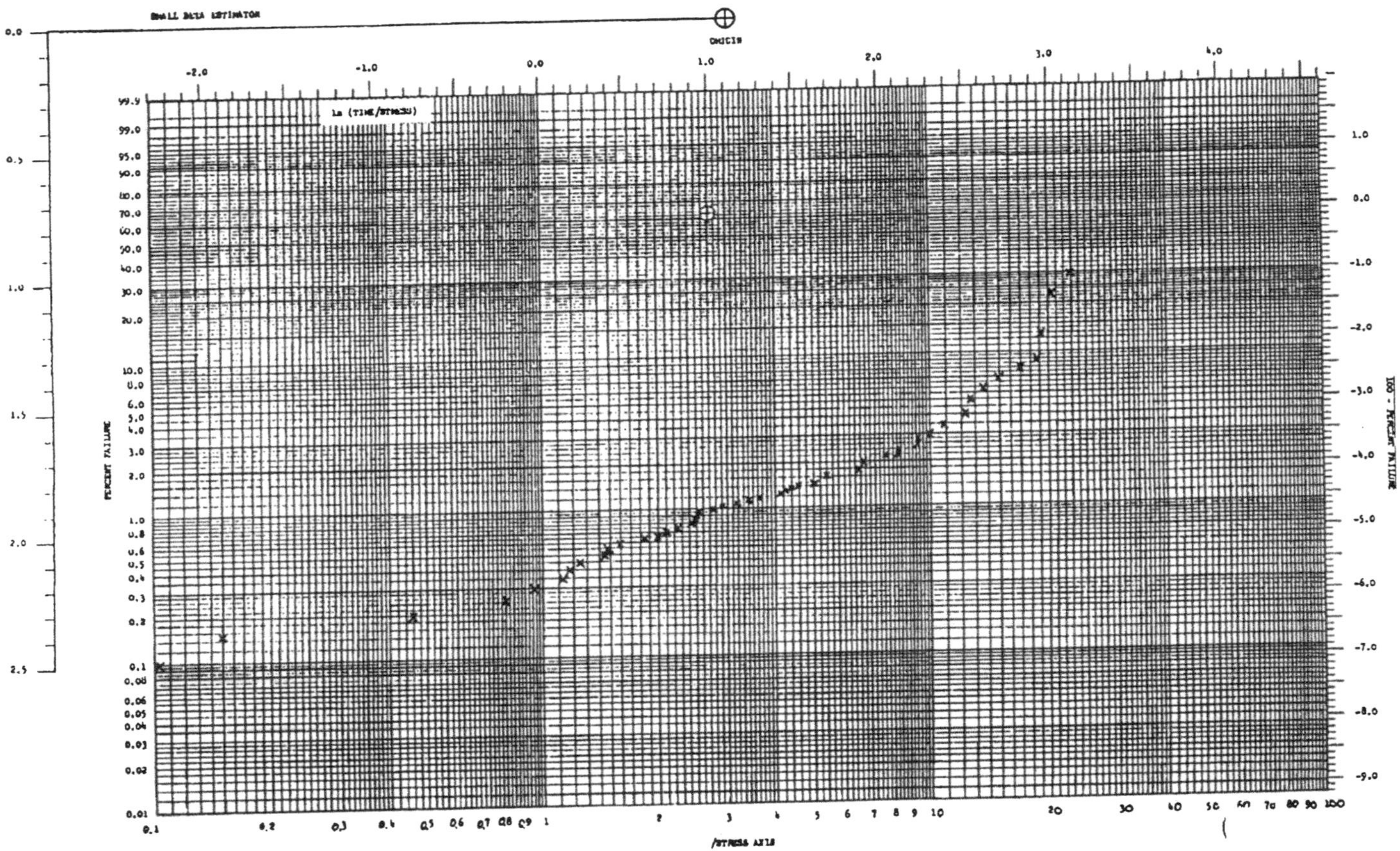

FIGURE 4.4. ***Pressure vessel failure data.***

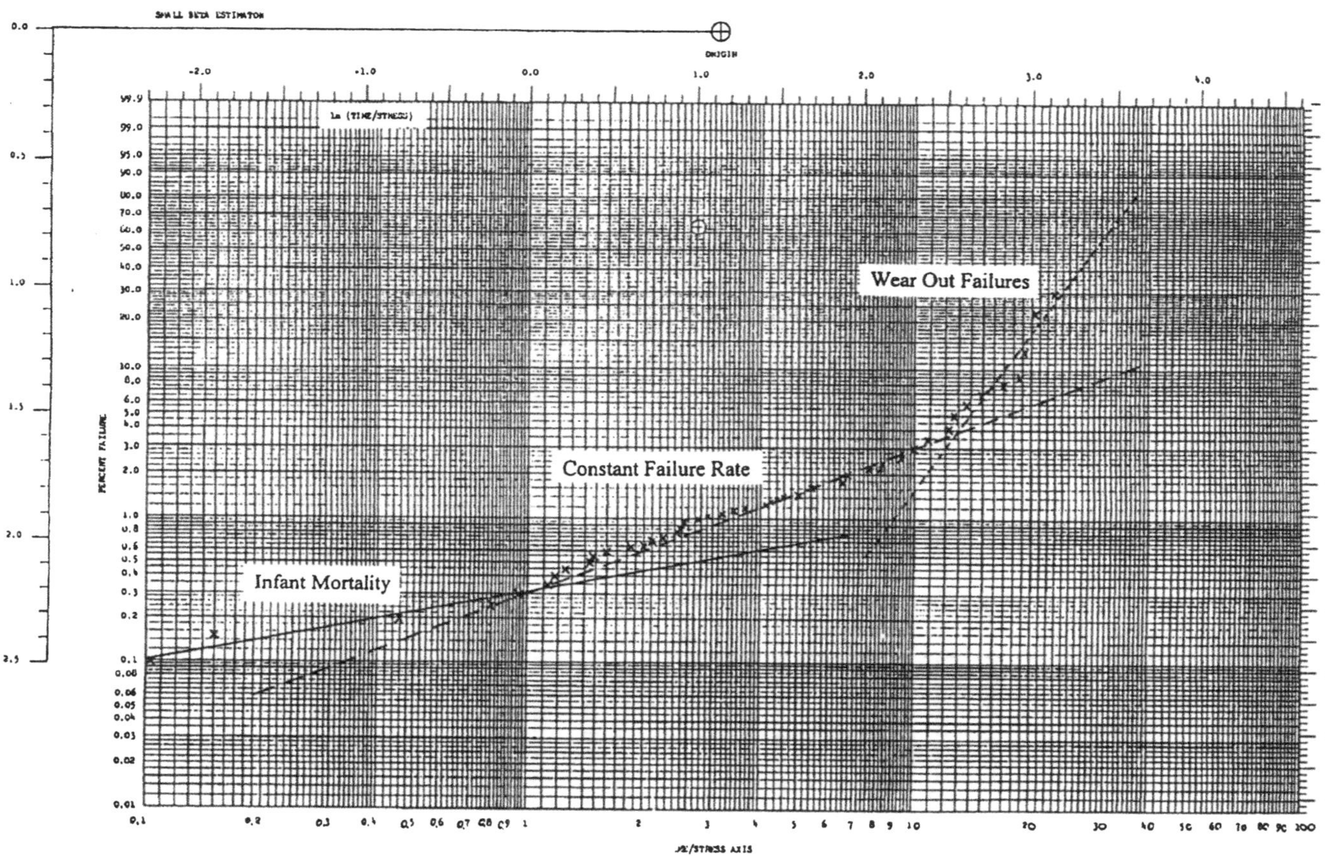

FIGURE 4.5. ***Weibull pressure vessel data plot.***

Note that these data seem to fall into three distinct failure distributions, as described approximately by the lines in Figure 4.5. The left side of the plot has a Weibull β of about 0.5, indicating an "infant mortality" failure distribution; that is, a falling failure rate. The middle of the plot has a β of about 1.0, indicating a constant failure rate, or exponential failure distribution. The final portion of the figure has a β of almost 3.5, suggesting a "wear out" portion of the equipment life where the failure rate is rapidly increasing with time.

The possible reasons for such behavior are varied, but common explanations are as follows:

- *Infant mortality*—Errors in design or construction; manifestation of unknown failure mechanisms, human errors resulting from lack of experience in operating the equipment. These types of problems tend to be realized shortly after startup, and then decline over time.
- *Constant failure rate*—Failures are random. It is useful to note that preventive maintenance during the constant failure rate regime may be counterproductive, in that maintenance activities, replacement parts, etc. may reintroduce infant mortality failures.
- *Wear out*—Failure mechanisms that are cumulative over time, such as corrosion, erosion, or fatigue. This accumulated degradation may become apparent near the end of the equipment item's design "useful life."

Incidentally, the failure probability density function $f(t)$ can be described, if desired. The Weibull form of this function is

$$f(t) = \frac{\beta}{\eta}\left(\frac{t}{\eta}\right)^{\beta-1} \exp\left[-\left(\frac{t}{\eta}\right)^{**} \beta\right]$$

The Weibull functions are then replotted in Figure 4.6. Note that this plot takes the form of the classic "bathtub" curve.

4.9. System Performance—Compressor Example

[*Note:* The examples in the following sections have been simplified in some respects compared to what an analyst in a plant environment might be required to undertake. For example, there is no consideration of present/future values of expenditures. The data analyst might also be more selective in choosing failure modes for consideration in some examples than is the case here (e.g., not including the "fail-to-start" mode in the failure *rate* calculation immediately below). Lastly, the term "failure" is used somewhat loosely in the following text. A "failure," in the

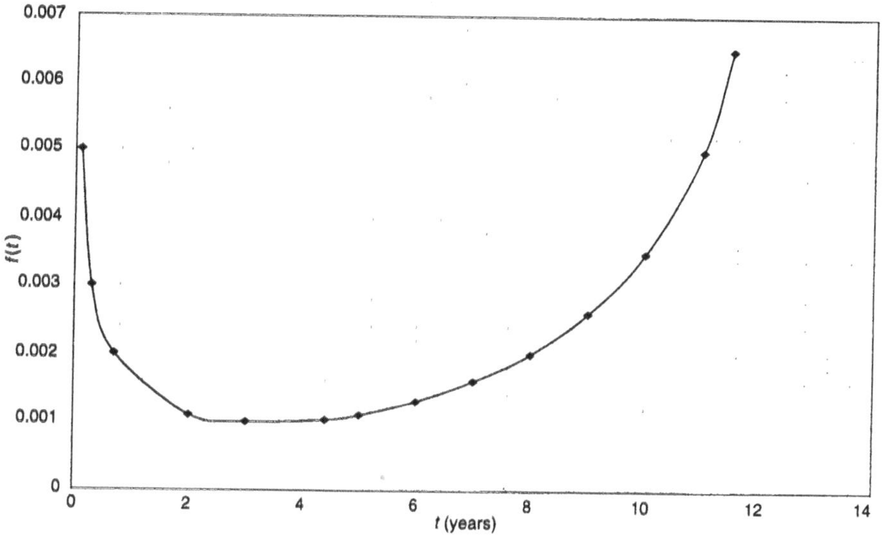

FIGURE 4.6. *"Bathtub failure distribution."*

context below, will certainly include events in which the equipment has entered a fault condition. But it may also include events such as planned maintenance which, although not resulting from a fault in that point in time, are still downtimes necessary to identify and correct pre-fault conditions.]

You have been asked to specify a centrifugal gas compressor for acid gas service. You have at your disposal the CCPS Database. You have received the process engineering data and recognize that the compressor system needs the following characteristics:

Capacity	3000 HP
Performance	less than 3 failures per year
Availability	98.5% or better

You were able to download the information on the following four pages from the Database on two brands of compressors that your company has experience using, and for which you have some spare parts. However, you have none in acid service and therefore would like to evaluate this equipment based on the experience of others participating in the Database.

ITEM: Compressor

FILTERS	APPLIED	Upper Level Inventories	Lower Level Inventories	External Environment	Internal Environment	Aggregated Time (10^6 hrs)	Population
			Centrifugal, Electric Motor Driven, 1000-4000 HP, Brand X	Outdoors, not enclosed	Acid gas	.2809	21
						Aggregated No. of Demands	**Contributing Sites**
						127	8

Selected Analysis: Event Rate

Units Of Measure: per 10^6 hours

Selected Confidence Limit: 90%

	Available Events	No. Of Events	Lower Estimate:	Mean	Upper Estimate
X	Fails to start	24	51.94	85.44	140.5
X	Fails while running	96	207.7	341.76	562.2
X	High discharge temperature	15	34.46	53.40	87.84
X	High vibration	9	16.71	32.04	55.91
X	Overhaul	12	24.65	42.72	69.22
	Sum of Selected Failure Modes	156		555.36	

Comments:

NA – Not applicable/available

ID – Insufficient data to support analysis

ITEM: Compressor

FILTERS	APPLIED	Upper Level Inventories	Lower Level Inventories	External Environment	Internal Environment	Aggregated Time (10^6 hrs)	Population
			Centrifugal, Electric Motor Driven, 1000-4000 HP, Brand X	Outdoors, not enclosed	Acid gas	.2809	21
						Aggregated No. of Demands	**Contributing Sites**
						127	8

Selected Analysis: Time to restore

Units Of Measure: hours

Selected Confidence Limit: None

	Available Events	No. Of Events	Lower Estimate:	Mean	Upper Estimate
X	Fails to start	24		5.8	
X	Fails while running	96		30.5	
X	High discharge temperature	15		20.6	
X	High vibration	9		15.4	
X	Overhaul	12		42.0	
Sum of Selected Failure Modes		156		25.79	

Comments:

NA – Not applicable/available

ID – Insufficient data to support analysis

ITEM: Compressor

FILTERS	APPLIED	Upper Level Inventories	Lower Level Inventories	External Environment	Internal Environment	Aggregated Time (10^6 hrs)	Population
			Centrifugal, Electric Motor Driven, 1000-4000 HP, Brand Y	Outdoors, not enclosed	Acid gas	.9604	61
						Aggregated No. of Demands	**Contributing Sites**
						402	15

Selected Analysis: Event rate

Units Of Measure: per 10^6 hours

Selected Confidence Limit: 90%

	Available Events	No. Of Events	Lower Estimate:	Mean	Upper Estimate
X	Fails to start	69	43.68	71.85	118.2
X	Fails while running	201	127.23	209.29	344.3
X	High discharge temperature	36	16.71	37.48	61.65
X	High vibration	21	13.29	21.87	35.98
X	Overhaul	15	9.50	15.62	25.69
	Sum of Selected Failure Modes	342		356.10	

Comments:

NA – Not applicable/available

ID – Insufficient data to support analysis

ITEM: Compressor

FILTERS	APPLIED	Upper Level Inventories	Lower Level Inventories	External Environment	Internal Environment	Aggregated Time (10^6 hrs)	Population
			Centrifugal, Electric Motor Driven, 1000-4000 HP, Brand Y	Outdoors, not enclosed	Acid gas	.9604	61
						Aggregated No. of Demands	**Contributing Sites**
						402	15

Selected Analysis: Time to restore
Selected Confidence Limit: None
Units Of Measure: hours

	Available Events	No. Of Events	Lower Estimate:	Mean	Upper Estimate
X	Fails to start	69		8.6	
X	Fails while running	201		44.8	
X	High discharge temperature	36		23.5	
X	High vibration	21		6.2	
X	Overhaul	15		68.0	
Sum of Selected Failure Modes		342		33.90	

Comments:

NA – Not applicable/available

ID – Insufficient data to support analysis

Check #1. Calculate the number of failures per year for each brand of compressor:

Brand X MTBF = 10^6 hours/555.36 = 1800.6 hours
(8760 hours/year)/(1800.6 hours/failure) = 4.87 failures/year

Brand Y MTBF = 10^6 hours/356.10 = 2808.2 hours
(8760 hours/year)/(2808.2 hours/failure) = 3.12 failures/year

Check #2. Calculate the average availability for each compressor:

Brand X 4.87 failures/year × 25.79 hours/failure
= 125.6 hours downtime/year
Availability = (8760 – 125.6)/8760 = 98.6%

Brand Y 3.12 failures/year × 33.9 hours/failure
= 105.8 hours downtime/year
Availability = (8760 – 105.8)/8760 = 98.8%

Therefore, both systems meet the specification for availability, but neither meets the goal for performance.

4.10. Life-Cycle Costs—Compressor Example (continued)

Your company has defined long-term cost of ownership as the sum of the following:

- Acquisition costs
- Operational costs
- Maintenance costs
- Disposal costs

After reviewing your application you have established the following costs of the types above. For each brand of compressor those costs are as follows:

	Brand X	*Brand Y*
Acquisition cost	$1,000,000	$1,250,000
Operational cost (running)	$3.50/hour	$3.50/hour
Operational losses (down)	$30,000/hour	$30,000/hour
Maintenance costs	$300/hour	$350/hour
Disposal costs	$10,000	$10,000

You have performed a "data dump," which shows that the actual resources used to complete work on the compressors are, on average, 51 man-hours and 72 man-hours for Brands X and Y, respectively. Calculate the long-term cost of ownership for each brand of compressor based on a service life of 10 years, and assuming no shutdowns for reasons other than compressor failure or overhaul.

Brand X

DOWNTIME: 555.36 failures/10^6 hours × 8760 hours/year × 10 years × 25.79 hours/failure = 1254.7 hours

DOWNTIME COSTS: 1254.7 hours × $30,000/hour = $37,641,000

RUNNING: (8760 hours/year × 10 years) – 1254.7 hours = 86345.3 hours

RUNNING COSTS: 86345.3 hours × $3.50/hour = $302,000

MAINTENANCE COSTS: 555.36 failures/10^6 hours × 8760 hours/year × 10 years × 51 hours/failure × $300/hour = $744,000

Brand Y

DOWNTIME: 356.10 failures/10^6 hours × 8760 hours/year × 10 years × 33.9 hours/failure = 1057.5 hours

DOWNTIME COSTS: 1057.5 hours × $30,000/hour = $31,725,000

RUNNING: (8760 hours/year × 10 years) – 1057.5 hours = 86542.5 hours

RUNNING COSTS: 86542.5 hours × $3.50/hour = $303,000

MAINTENANCE COSTS: 356.10 failures/10^6 hours × 8760 hours/year × 10 years × 72 hours/failure × $350/hour = $786,000

10-Year Cost of Ownership	*Brand X*	*Brand Y*
Acquisition Cost	$ 1,000,000	$ 1,250,000
Operational Cost (running)	$ 302,000	$ 303,000
Operational Cost (down)	$37,641,000	$31,725,000
Maintenance Cost	$ 744,000	$ 786,000
Disposal Cost	$ 10,000	$ 10,000
TOTAL	$39,697,000	$34,074,000

4.11. Maintenance Budgeting—Compressor Example (continued)

Your maintenance manager would like your input into what the expected resources would be for each brand of equipment for overhauls as well as how long the overhauls might take. You have found from a second data dump that an overhaul for Brand X requires 86 man-hours and an overhaul takes 142 man-hours for Brand Y.

Although there appears to be a significant long-term cost of ownership difference favoring Brand Y, the maintenance manager is concerned about the higher cost of maintenance per hour for Brand Y. Calculate the cost of overhauls over the life expectancy of this equipment.

Brand X
(42.72 overhauls/10^6 hours) × 8760 hours/year × 10 years = 3.68
~ 4 overhauls
4 overhauls × 86 hours/overhaul × $300/hour = $103,200

Brand Y
(15.62 overhauls/10^6 hours) × 8760 hours/year × 10 years = 1.37
~ 2 overhauls (being conservative)
2 overhauls × 142 hours/overhaul × $350/hour = $99,400

Thus the costs for overhauls are about the same based on the expected improved failure rate for Brand Y, although its overhaul time is greater and the unit rate time is greater.

4.12. Throughput Targets—Compressor Example (continued)

You have been asked to evaluate the system throughput for the life of the project, with an annual production goal of 2,600,000,000 SCF of acid gas. You are given three alternative design configurations using either Brand X or Brand Y:

1. two compressors rated at 50% system capacity (1500 HP/2500 SCFM)
2. two compressors rated at 60% system capacity (1800 HP/3000 SCFM)
3. two compressors rated at 100% system capacity (3000 HP/5000 SCFM)

In Alternative 1, when one compressor is shut down, production is reduced to 50% capacity. In Alternative 2, when one compressor is shut down the operating conditions will be adjusted so that the compressor remaining on-line will carry 60% capacity. In Alternative 3, the system has full redundancy.

Assumptions

1. Only one compressor will be shut down at a time
2. Alternative 3 assumes that once the second compressor is started, a perfect switchover can take place—that is, there is no probability for the system failing to start
3. There is an average of two "zero time" shutdowns per year for each type of compressor for reasons other than failure or overhaul (power blip, unit trip, etc.)

Alternative Design 1, Brand X:

Number of failures other than failure to start
= 2 compressors × [(555.36 – 85.44)/10^6] × 8760 hours/year × 10 years
= 82.33 failures

Average number of hours per shutdown mode other than failure to start
= [(30.5)(96) + (20.6)(15) + (15.4)(9) + (42)(12)]/[96 + 15 + 9 + 12]
= 29.39 hours/shutdown

Number of failures to start in 10-year period
= (24 fails to start/127 demands)
× [(82.33 + (2 comp.) × (2 fails/year) × (10 years)]
= 23.12 failures to start

Average number of hours per failure to start shutdown mode
= 5.8 hours/shutdown

Number of hours in single compressor mode in a 10-year period
= (23.12 × 5.8) + (82.33 × 29.39)
= 2554 hours

Number of hours in double compressor mode in a 10-year period
= (8760 hours/year × 10 years) – 2554 hours
= 85046 hours

Total throughput in 10-year period
= 2554 hours @ 2500 SCFM × 60 minutes/hour
+ 85046 hours @ 5000 SCFM × 60 minutes/hour
= 25,897,000,000 SCF

Annual average throughput = 2,590,000,000 SCF

Alternative Design 2, Brand X:

Total throughput in 10-year period
= 2554 hours @ 3000 SCFM × 60 minutes/hour
+ 85046 hours @ 5000 SCFM × 60 minutes/hour
= 25,974,000,000 SCF

Annual average throughput = 2,597,000,000 SCF

Alternative Design 3, Brand X:

8760 hours/year × 10 years × 60 minutes/hour × 5000 SCFM
= 26,280,000,000 SCF

Annual average throughput = 2,628,000,000 SCF

Alternative Design 1, Brand Y:

Number of failures other than failure to start
= 2 compressors × [(356.10-71.85)/10^6] × 8760 hours/year × 10 years
= 49.80 failures

Average number of hours per shutdown mode other than failure to start
= [(44.8)(201) + (23.5)(36) + (6.2)(21) + (68)(15)]/[201 + 36 + 21 + 15]
= 40.30 hours/shutdown

Number of failures to start in 10-year period
= (69 fails to start/402 demands)
× [(49.80 + (2 comp.) × (2 fails/year) × (10 years)]
= 15.41 failures to start

Average number of hours per failure to start shutdown mode
= 8.6 hours/shutdown

Number of hours in single compressor mode in a 10-year period
= (15.41 × 8.6) + (49.80 × 40.30)
= 2139 hours

Number of hours in double compressor mode in a 10-year period
= (8760 hours/year × 10 years) – 2139 hours
= 85461 hours

Total throughput in 10-year period
= 2139 hours @ 2500 SCFM × 60 minutes/hour
+ 85461 hours @ 5000 SCFM × 60 minutes/hour
= 25,959,000,000 SCF

Annual average throughput = 2,596,000,000 SCF

Alternative Design 2, Brand Y:

Total throughput in 10-year period
= 2139 hours @ 3000 SCFM × 60 minutes/hour
+ 85461 hours @ 5000 SCFM × 60 minutes/hour
= 26,023,000,000 SCF

Annual average throughput = 2,602,000,000 SCF

Alternative Design 3, Brand Y:

8760 hours/year × 10 years × 60 minutes/hour × 5000 SCFM
= 26,280,000,000 SCF

Annual average throughput = 2,628,000,000 SCF

4.13. Summary

As with many topics in the area of statistics, the potential complexity of equipment failure applications is limited only by the imagination and fortitude of the analyst. The applications suggested above are relatively simple illustrations of some common uses of failure data. Many details regarding data confidence, possibly better alternative failure rate distribution descriptions, plant economics and many other topics have not been discussed, or have been greatly simplified. A number of simplifying assumptions have also been made.

It is simply not possible for this book to address all possible approaches in which plant personnel might be interested, and in general it is suspected that the types and sophistication of most applications will not be greatly different than those outlined above.

Nonetheless, the more adventurous are invited to pursue these applications at a higher level. Following are some suggested texts.

References

1. R. B. Abernathy, "The New Weibull Handbook, 2nd ed.," distributed by Gulf Publishing, Houston, TX, 1996.
2. H. P. Bloch and F. K. Geitner, "An Introduction to Machinery Reliability Assessment, 2nd ed.," Gulf Publishing, Houston, TX, 1994.
3. J. Davidson, "The Reliability of Mechanical Systems," Mechanical Engineering Publications Ltd., for the Institution of Mechanical Engineers, London, 1988.

4. R. A. Dovich, "Reliability Statistics," ASQC Quality Press, Milwaukee, 1990.
5. D. J. Smith, "Reliability and Maintainability in Perspective," 3rd ed., Macmillan Education Ltd., London.

5 Data Structure

5.1. Data Structure Overview

This chapter describes the structure of the database established to collect data and perform calculations described in the previous sections of this Guideline. This structure is embodied in database software that accepts and stores data, analyzes the data according to predefined algorithms and rules sets, and produces a set of standardized results for the user.

A summary of the data structure for the CCPS Equipment Reliability Database is shown in Figure 5.1. This figure is valid (and can be applied) at any level in the taxonomy. The CCPS Database contains four distinct types of data:

- *Unique Item Data*—Fixed data specific to an item as defined by an identifying number and a boundary diagram. Examples of unique item data are serial number, model number and manufacturer.
- *Application Data*—User-defined data specific to the location and the condition of operation and maintenance of an item. Examples of application data are pressure, temperature, fluid state, and corrosiveness.
- *Event Data*—Data recording events that can happen to, or are done to a given item in a given application, for example, inspections, tests, or failures. An example of an event datum would be the actual lift pressure from a relief valve bench test.
- *Failure Logic Data*—For failure events, these are data that define the item function(s) lost, the mode(s) of failure observed and the mechanism of failure. For example, a failure logic datum might specify that a lift pressure of more than 150% of the design setpoint would translate to the failure mode of "fails to lift at set pressure."

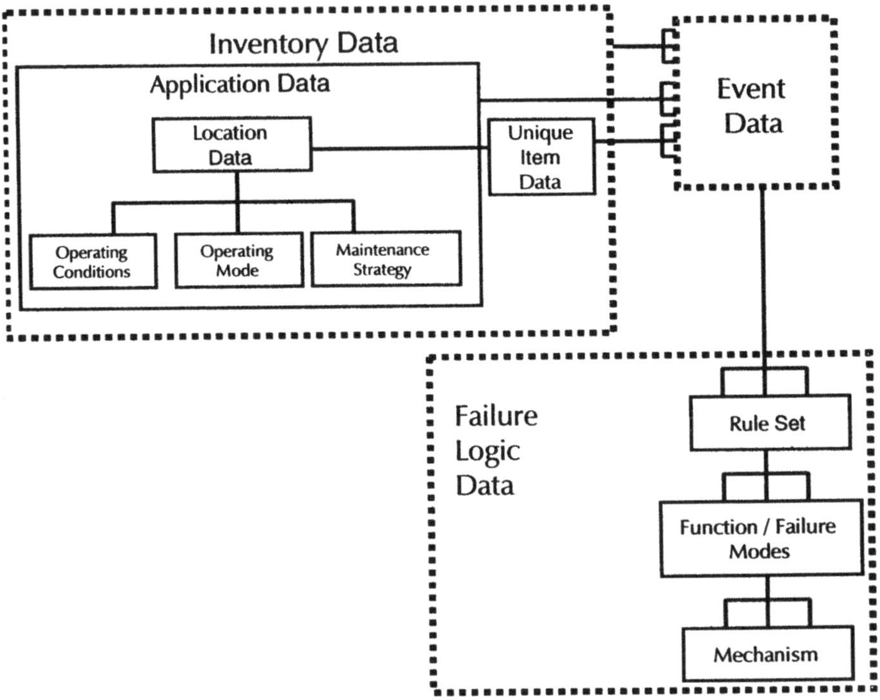

FIGURE 5.1. ***Data structure overview.***

Note that, within the Inventory Data group, the CCPS Database makes a distinction between information relating to the application and the information relating to a particular piece of equipment. This distinction is made because a company may utilize a piece of equipment in several different locations during the life of that equipment item. The structure allows the database user to collect and analyze information for both location and item.

5.2. General Taxonomy

The general taxonomy for the Application and Inventory Data of the CCPS Database is shown in Figure 5.2. Because the Application Data are user-defined and "boilerplate" in nature, it is addressed first in the taxonomy. The General Taxonomy has seven basic levels of indenture.

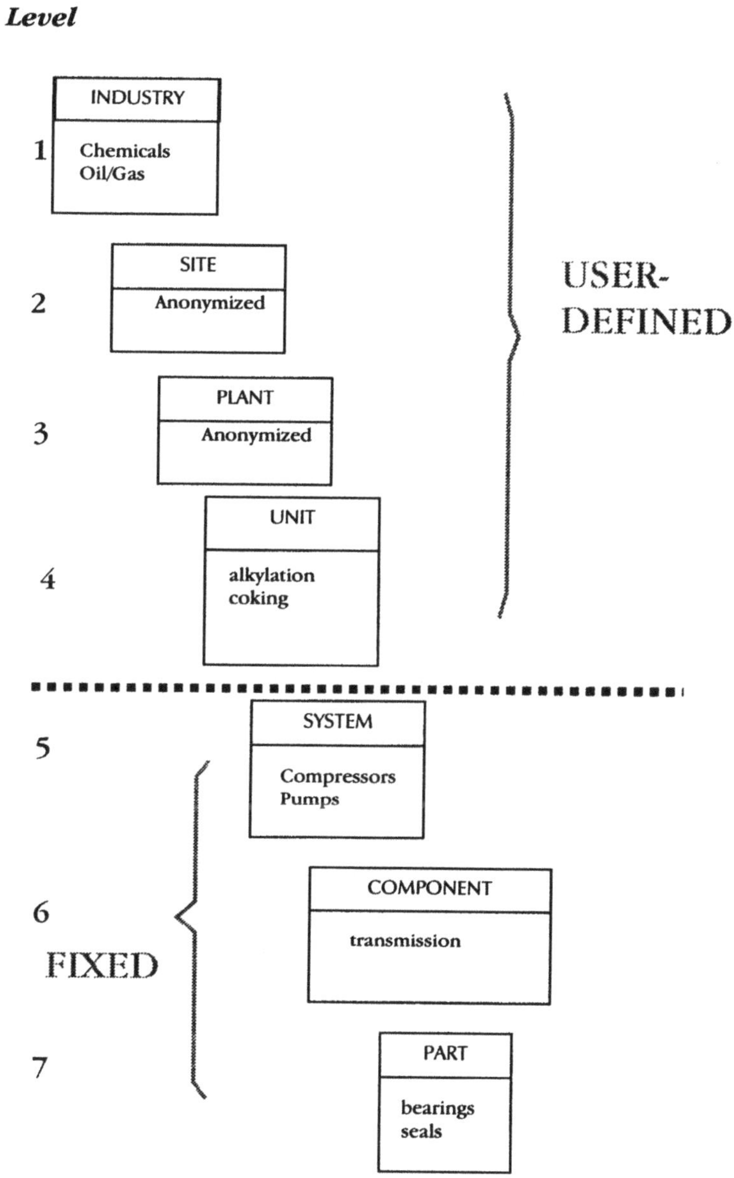

FIGURE 5.2. ***CCPS Database taxonomy.***

1. Industry
2. Site
3. Plant
4. Unit / Process Unit
5. System
6. Component
7. Part

The first four levels of the taxonomy relate to application data and the last three relate to inventory data, comprising both item and application information. The options for all levels are listed in Appendix II.

It is not practical to display the entire taxonomy on paper. It is better to show the basic relationships that exist within the software, supplemented by standard pick list tables and failure mode definitions. It is possible for the user to query the relational database using filters, to essentially define a unique taxonomy with attendant data for the purpose of analyzing specific issue(s).

5.2.1. Taxonomy Levels 1–4 (Industry, Site, Plant, Process Units)

The taxonomy appearing at levels 1–4, as shown in Figure 5.2, is a high level categorization that relates to industries and plant operations regardless of the equipment systems involved. For these levels, the user inputs the appropriate choices from pick lists for each equipment system being tracked. In this way the actual taxonomy for the specific application is developed. During analysis the user may query the data to define a special taxonomy of interest. This is illustrated in Table 5.1.

TABLE 5.1
Illustration of User-Defined Taxonomy

Level	Industry	Site	Plant	Unit	System
Pick List	Chemical—commodity	Ohio	Refinery	Alkylation	Pumps
	Chemical—polymers	New York	Olefins	Coking	Compressors
	Gases—industrial	Texas A	Ethylene	Distillation	Turbines
	Oil/Gas—refining	Texas B		Reforming	Vessels
				Hydrocracking	Condenser

The query above would search for data for all pumps in distillation units in the refinery plant at the 'Texas A' site, within the oil–gas refining industry. The anonymization process means that only users with specific privileges would be able to specify searches for their own site. An example of the complete taxonomy, down to the part level, might be

Level 1/Industry — chemical
Level 2/Site — Texas A
Level 3/Plant — ethylene
Level 4/Unit — compression
Level 5/System — compressor
Level 6/Component — power transmission
Level 7/Part — coupling

The above example assumes that the user has access rights to specific site and plant data. For users without access to the proprietary fields in the data, those fields cannot be specified in a search query, and are bypassed in the taxonomy. An example of a typical search by someone with no access rights might be as shown below:

Level 1/Industry — chemical
Level 2/Site — all
Level 3/Plant — all
Level 4/Unit — compression
Level 5/System — compressor
Level 6/Component — power transmission
Level 7/Part — coupling

5.2.2. Taxonomy Levels 5–7 (System, Component, Part)

The taxonomy of levels 5–7 are specific to equipment systems. Use of the relational database software allows the user to specify equipment system populations that are of specific interest to the analyst. In one case, the analyst may be interested in compressor systems processing a particular fluid. For this case, all equipment considered part of the compressor would be included. In another case, the concern may be the performance of heat exchangers. Only heat exchangers of a particular type, such as shell and tube, would be included for this case.

The equipment systems available within the CCPS database are listed in Appendix II. For the convenience of the user, equipment systems have generally been grouped into various classes of systems. The concept of "grouping" was borrowed from the ISO-standard and the OREDA database. "Equipment Group" provides a convenient means for grouping equipment systems.

5.2.3. *Treatment of Subordinate Systems in the CCPS Database*

Some systems, such as heat exchangers, function as independent systems in some applications or as support systems in other applications. In a supporting role, these systems act like components of the larger system, rather than independent systems. These supporting systems are referred to as ***subordinate systems*** when they function as components for a larger system. The subordinate relationship between systems is sometimes referred to as a "parent–child" relationship. An example of this parent–child relationship is the interstage heat exchanger that is part of a compressor system. The compressor is the parent, and the heat exchanger is the child. However, both the compressor and the heat exchanger are equipment systems in the taxonomy.

To resolve this dilemma, the Equipment Reliability Database has included the option to record the subordination of systems, so that failure of a subordinate system accurately cascades to failure of the primary system. The systems are registered as independent systems, but they are tagged as being subordinate to a larger system. Using the example of a heat exchanger supporting a compressor, the user would enter the database in the compressor system and would then be automatically linked to the heat exchanger system. The heat exchanger would appear as a component when system-level events are recorded.

5.3. Database Structure

The database is constructed of a series of tables that specify the database fields. Software is also available to analyze and report data according to a user's specifications. Each table contains numerous fields that are populated with predefined information, known as "pick lists." The pick lists provide a means of standardizing input fields (see Appendix II for the CCPS Database tables).

Table 5.2 shows an overview of the tables contained in the database. During future development, more tables may be added. Temporary tables, used to store intermediate results for calculations, are not shown in Table 5.2.

Descriptions of the tables used in the database are provided in the following sections. Due to the changing nature of the database design as additional equipment systems are developed and added, the exact tables are not shown in this Guideline.

5.3.1. Inventory Tables

Inventory tables provide information that identifies a piece of equipment for use in the database. Inventory information is used when filtering data for specific applications.

Subscriber Inventory

Data are submitted to the database by a Subscriber. Data in this field are fully anonymized in the aggregated dataset distributed to participants. Typically, the Subscriber would be based at a plant or site and be registered and qualified to submit data. It is possible for one company to have several subscribers to the database. There is one copy of the Subscriber Inventory Table for each Subscriber. Throughout the data tables, the Subscriber is uniquely defined by the Subscriber Name. The Subscriber Inventory Table captures information about the Subscriber such as company name, contact name, address (physical, postal, and e-mail), and phone numbers.

Site Inventory

Data in this field are fully anonymized in the aggregated dataset distributed to participants. This table is completed once for each site enrolled by a subscriber in the reporting system. There is only one copy of this table for each site. In most cases, there will only be one site associated with a subscriber. In many cases, depending on the internal organization of the reporting entity, the site name will be the same as the Subscriber Location. This table contains information such as site name, location, and contact.

Plant Inventory

This table is completed once for each Plant in the reporting system at a given Site. Data in this field are partially anonymized in the aggregated dataset distributed to participants. There may be multiple versions of this table for each site. This table captures information such as plant name, industry associated with the plant, as well as type of plant, external environment, plant start date, and normal operating mode.

Unit Inventory

This table is submitted once for each process unit in the reporting system at a plant. There may be multiple versions of this table for each plant. The table contains information regarding unit name, operating mode, and unit start date.

TABLE 5.2
Database Structure

Table Type	Table Description	Related Tables	Notes
Inventory Tables	Subscriber Inventory Site Inventory Plant Inventory Unit Inventory		
	System Inventory—General	System-Specific Inventory	System-specific fields are a function of system type
Event Tables	Plant Events	Loss of Containment Startup Shutdown Curtailment	
	Process Unit Events	Loss of Containment Startup Shutdown Curtailment Calibration Inspection Proof Test Repair Preventive Maintenance Condition-Based Maintenance Demand	
	System Events	Loss of Containment Startup Shutdown Curtailment Calibration Inspection Proof Test Repair Preventive Maintenance Condition-Based Maintenance Demand	Not all event types are Applicable to all Systems.

System Inventory—General

There is one General System Inventory table for all systems enrolled in the database, regardless of Subscriber. There is one record for each system logged in the database. Each record in the General Inventory table captures information that applies to all systems, regardless of type. A sample of some of the fields in this table is shown below.

- Subscriber ID
- Site ID
- Plant ID
- Serial number
- Tag number
- Date installed
- Surveillance start date
- Maintenance strategy
- Associated systems
- External environment

System Inventory—Specific

Much of the information needed to store meaningful inventory data is a function of the specific equipment type. For instance, a relief valve can be partially characterized by its lift pressure, a property that a pipe would not have. Because different types of systems require different fields to specify the inventory, a specific inventory table exists for each type of system in the CCPS Database. As an example, a partial listing of the System-Specific Inventory table for relief devices is provided below:

Valve design
- Manufacturer
- Model number
- Inlet connection size
- Orifice size and designation
- Outlet connection size

Service Conditions
- Fluid handled
- Phase at relief inlet
- Set/Opening pressure
- Blowdown %

Selection of material
- Body material
- Spring material
- Bonnet material
- Seat material

5.3.2. Event Tables

Once an item (plant, unit, system, etc.) has been registered in the database, event information can be gathered via the event tables. Currently the CCPS Database is designed to record data at the plant and system levels.

A key feature of the CCPS Equipment Reliability Database is the capacity to store and retrieve data pertaining to events at the plant, process unit and the system levels.

Plant Events

The main Plant Event table records the name and type of event occurring for a specific plant registered in the database. The information stored in the Plant Event table is the name of the plant, date and time of the event and the type of event (curtailment, shutdown, etc.). Once the type of event has been recorded, the user inputs specific data through the subtables listed below.

Plant Event—Loss of Containment

This table stores information for loss of containment events at the Plant level. Some of the fields in this table are listed below.

- Event date and time
- Event duration OR
- Event end date and time
- Reportable? Y/N
- Fluid released
- Failure descriptors
- Severity
- Hole size
- Consequence

For loss of containment events, the severity field is then used to determine the specific failure mode based on a rule set programmed or imbedded in the database program.

Plant Event—Shutdown

The Plant Shutdown Event table records necessary information for a plant shutdown. A partial listing of the fields in this table is shown below.

- Event date and time
- Event duration or
- Date and time returned to service
- Is there a *force majeure*? Y/N

- Shutdown cause
- Shutdown subcause
- Scheduled activity? Y/N/Unknown

The default for all plant shutdown events is "loss of production." Through the use of filters, an analyst can zero in on why the plant shut down.

Plant Event—Startup
The Startup Event table for plants contains the following fields:

- Event duration
- Event end date
- Event end time
- Was the startup successful? Y/N

Plant Event – Curtailment
A curtailment is defined as a partial reduction in the output of a plant. Curtailment information is gathered using the following fields:

- Event start date and time
- Event duration, or
- Date and time returned to full capacity
- Cause of curtailment
- Maintenance Required?
- Average percent capacity available

Process Unit Events

Process unit events had not been developed at the time of publication of this Guideline.

System Events

The events below are examples only. Some fields and/or tables may change based upon system of interest. The System Events table contains the records submitted by a subscriber for each system event reported during a given period. One record is submitted for each event. In the CCPS Database, the following are considered as events for systems:

- Loss of Containment
- Startup
- Shutdown
- Curtailment
- Calibration

- Inspection
- Proof Test
- Repair
- Preventive Maintenance
- Condition-Based Maintenance
- Process Demand

Owing to the differences among systems, some of the above events may not apply to a given system. For instance, a curtailment is not an applicable event for relief valves. When the taxonomy for a system is established, the list of applicable events is developed so that the database can be programmed to screen out irrelevant events.

Any number of records can be reported during a given period. All events are uniquely defined by their Subscriber ID, System ID, event date and time.

System Event—Loss of Containment

Loss of containment events can occur for nearly all systems. A sample of the fields used for system-level loss of containment events is shown below for relief valves:

- Event date and time
- Event end date and time or
- Duration
- Pressure at time of failure
- Fluid Released
- Severity
- Quantity Released, lb
- Hole Size
- Estimated Area
- Event description
- Consequence
- Reportable Incident?
- If reportable, reportable to:
- Comments

System Event—Shutdown

If a system is shutdown, certain data are recorded for the estimation of relibility parameters.

- Event date
- Event time
- Date and time returned to service

- Is there a *force majeure?* Y/N
- Shutdown cause
- Shutdown subcause
- Scheduled activity? Y/N/Unknown

System Event—Startup

The system startup event is characterized by the ability of the system to start according to predefined criteria. A failure to start is, by itself, a failure mode for most systems. If the field for successful start is a no (N), the failure mode is automatically set to "fails to start."

System Event—Curtailment

Curtailments events are usually linked to the "partial loss of throughput" failure mode. The following is a sample of fields appearing in a System Curtailment table.

- Event start date and time
- Event duration, or
- Date and time returned to full capacity
- Cause of curtailment
- Maintenance Required?
- Average percent capacity available

System Event—Inspection

Inspections typically do not mean that a failure has occurred; however, they can point to a failure mode that may not be easily detected. The Inspection Event table records the necessary information from a system inspection. For relief valves, a sample of the data fields are provided below.

- Discharge piping visual inspection
- Weep hole visual inspection
- Evidence of seat leakage
- External leakage
- Repair required
- Pressure between rupture disk and relief valve
- Insulation loss?
- Heat tracing operating?
- Pressure monitor functional?
- Temperature monitor functional?
- Excess flow check valve functional?

System Event—Process Demand

Process upsets may trigger failures or successes for some systems. A partial listing of the process upset data collected for relief devices is provided in the list below.

- Event end date
- Event duration or event end time (Depending what is picked, the other should be calculated.)
- Reason for process demand
- Event description
- Valve lift? (Discharge through the relief valve.)
- Lift pressure
- Maximum process pressure achieved
- Rapid cycling of the valve (chatter) audible during incident?
- Valve reseat?
- Reseat pressure
- Code deviation?
- Protected equipment damaged?
- Loss of containment?
- Repair or replacement required?
- Additional information

System Event—Repair

Repair events occur when a repair is made following a failure, proof test, inspection, or condition-based maintenance. The following fields are part of the System Repair Event table for relief devices. Keep in mind that "failure mode" does not appear as a field here, since failure modes are calculated by the rule set provided to the software.

- Maintenance start date
- Maintenance start time
- Date maintenance completed
- Maintenance hours expended
- Maintenance performed
- Components repaired
- Components replaced
- Part replaced
- Components failed
- Component failure cause(s)
- Part failure mode
- Part failure cause(s)
- Additional information
- Disposition following repair

System Event—Proof Test

The fields used to record data for a relief device proof test event are shown below.

- Proof test date
- Event time
- Visual flow path inspection: inlet
- Visual flow path inspection: outlet
- External leakage
- Seat leakage
- Leak pressure
- Valve lift?
- Lift pressure
- Maximum test pressure
- Valve reseat
- Reseat pressure
- Leakage following reseat
- Maintenance performed?
- Actual test time (hours)
- Additional information

System Event—Calibration

The fields used to record data for a calibration are shown below.

- Calibration date
- Time of calibration
- Calibration performed in-situ (Y/N)
- As-found value(s)
- Point check (P) or range (R)?

System Event—Condition-Based Maintenance

Condition-based maintenance (CBM) events occur when maintenance is performed based on the some predefined condition of the item. The following fields are part of the System CBM Event table for relief devices.

- Maintenance start date
- Maintenance start time
- Condition for maintenance
- Maintenance hours expended
- Maintenance performed
- Components repaired
- Components replaced
- Part replaced

- Components failed
- Component failure cause(s)
- Part failure mode
- Part failure cause(s)
- Additional information
- Disposition following repair

System Event—Preventive Maintenance
Preventive maintenance events occur when maintenance is carried out at predetermined intervals or according to prescribed criteria and is intended to reduce the probability of failure of an item. The following fields are examples of the Preventive Maintenance Event table for relief devices.

- Maintenance start date
- Maintenance start time
- Date maintenance completed
- Maintenance hours expended
- Maintenance performed
- Components repaired
- Components replaced
- Part replaced
- Components failed
- Component failure cause(s)
- Part failure mode
- Part failure cause(s)
- Additional information

5.3.3. Failure Logic Data

Failure logic data are not contained in separate tables; rather, they are embedded in the programmable logic of the database, allowing the failure modes to be determined from factual data contained in event data tables. The failure logic is determined during the taxonomy development process and provides the crucial link between event data and failure modes. Samples of the failure logic data for relief devices are presented in Table 5.3.

TABLE 5.3
Sample of Failure Logic for Relief Valves

System Failure Modes	Failure Mode Definition	Relief Valve Failure Logic	
		Input Form	**Criteria**
Fail to open	Fail to open prior to reaching 1.5 times set pressure	Demand	Relief lift = "NO" or lift ratio equal to or greater than 1.5
		Proof Test	Fail to open if "plugged" or lift ratio equal to or greater than 1.5 or max pressure equal to or greater than 1.5
		Inspection	Discharge piping plugged field = Plugged
Opens above set pressure	Opens between 1.1 and 1.5 times set pressure	Demand	Lift ratio between 1.1 and 1.5
		Proof Test	Lift ratio between 1.1 and 1.5

6 Quality Assurance of Data

6.1. Introduction

Any analysis requiring equipment reliability data is strongly dependent on the quality of the data available. Data are judged to be of high quality if they have the following characteristics:

- They are complete with respect to the information necessary for the intended analysis.
- They are gathered in compliance with a well documented set of definitions.
- They are accurately input, and the data are handled using quality principles.

The purposes of quality assurance are to specify the quality objectives and the requirements for the CCPS Industry Aggregated and Anonymized Database, verify that these requirements are being fulfilled, and to specify a process for handling deviations from the requirements.

This chapter contains the basic guidelines for establishing a management system for defining and verifying the quality of the data collection process. It provides the concepts, not the details. Many of the details are provided in Appendix I; others are provided as part of the Database operating instructions, which are evolving.

Note: *The discussion that follows is what has been agreed to by members of the CCPS Database effort. These apply to this database effort only, and may be more prescriptive than one would choose for an internal company effort. The CCPS Database group is combining data from a large number of sources, and therefore has inherently greater potential for inconsistent standards for data collection and reporting. The methods detailed in what follows are what the combined Database members have agreed are necessary to achieve the goals of this particular database.*

6.2. Basic Principles of Quality as Applied to Equipment Reliability Data

Any quality work process should meet the fundamental objectives of a quality management—it should fulfill the cycle of "plan–do–check–act" for achieving objectives through continuous improvement. Putting the management cycle into practice leads to seven major components in the quality system:

1. The data collection work process should have a **documented mission** defining the database's purpose and expectations.
2. The **basis** for the collection work process design, its expected output, and use of output should be clearly specified.
3. A **scope**, clearly defined, specifying inputs/outputs, resource requirements and limitations is required.
4. A **database administrator** should be appointed as responsible for ensuring that the process performs as expected over time.
5. The work process should specify **communications** of quality system results, specifying to whom, how often, and by what means.
6. A **verification system**, including a measurement process to assess whether or not objectives are being met and communicated to members is required in the quality plan for the data collection process
7. A **continuous improvement** mechanism should be in place to evaluate and implement changes to the database operating concept.

Specific guidance to assist the development of the data quality work process is contained in the following sections.

Quality Process Objectives

The quality process should meet the performance criteria of existing standardized data collection standards such as ISO 14 224 "Collection of Reliability and Maintenance Data for Equipment." Appendix I of this document further describes principles in collecting quality data, including the characteristics of quality data mentioned at the start of Section 6.1 of this book.

The ISO standard specifies the following precursors to data collection:

- Verifying completeness of data sources
- Having defined objectives, to make sure the proper information is collected
- Verifying that data sources have the proper information

- Checking that basic knowledge (e.g. installation date, population, operating period) on the equipment is available
- A pilot test of the data collection process (suggested)
- A data collection process (such as schedules, milestones, etc.) should be specified
- Training, motivation and organization of personnel should be in place
- A plan for quality assurance of the data collection process should be in place

6.3. Quality Management

A management system designed to ensure quality is essential for both the data subscribers/data contributors and the Database administrator. The quality work process applies to the data, software, and any publications generated by information from the Database. The quality assurance system for the Database itself must be approved and implemented prior to the acceptance of any data into the Database. This quality requirement does not apply to beta-testing or start-up testing of the Database, however it does require purging the system of this information prior to actual operation.

The data quality work process is intended to function for data that are manually input or retrieved by software translation and transfer from other management information systems.

6.4. Quality Principles

The quality assurance of data is to be an integral part of all major activities in the data collection and analysis process. Although the details of the quality plan are not specified in this chapter, some of the following basic requirements will be contained in the Database quality work process:

1. Policies shall be developed and implemented for the data validation process. Only validated data shall be accepted into the database.
2. A data subscriber will be named as a single point of contact in each quality plan. This contact person may or may not be responsible for data submission from more than one plant (data contributor).
3. Database participants must approve all required fields.
4. Data shall be supplied for all required fields.

5. Data shall be recorded according to standard approved and implemented procedures.
6. A method to ensure data security must be in place.
7. A forum shall be provided for promoting a common interpretation and understanding among data subscribers, including an exchange of experience.
8. All deviations shall be formally documented and submitted in a timely manner.
9. Quality review meetings shall occur on a periodic basis.
10. Database quality work process reports shall be made on an annual basis, summarizing major deviations, corrective actions, and changes to the quality work process.

Historical Data

From the perspective of the CCPS Database, historical data must adhere to the same standard of quality that is expected of "contemporary" data. This may result in the inability to use large amounts of accumulated information. However, the CCPS Database is more concerned about the effects of populating the Database with large amounts of uncontrolled data than the loss of such data.

6.5. Verification of Data Quality

The quality of data in the CCPS Equipment Reliability Database is ensured by a combination of documents and activities:

1. A certification process for the data subscriber
2. A quality plan administered by the data subscriber
3. Internal verification of data quality by data subscriber/data contributor
4. A quality plan administered by the CCPS Database administrator
5. Verification of data by the CCPS Database administrator prior to acceptance into the Database
6. External audit of the quality assurance work process in place

Figure 6.1 depicts the flow of data through the quality system in the CCPS Database, from acquisition to final inclusion in the database. Note: The data collection/conversion step can be either manual or electronic, although electronic is preferred. The data produced by the data conversion process are fed into the CCPS software that is supplied to the participants.

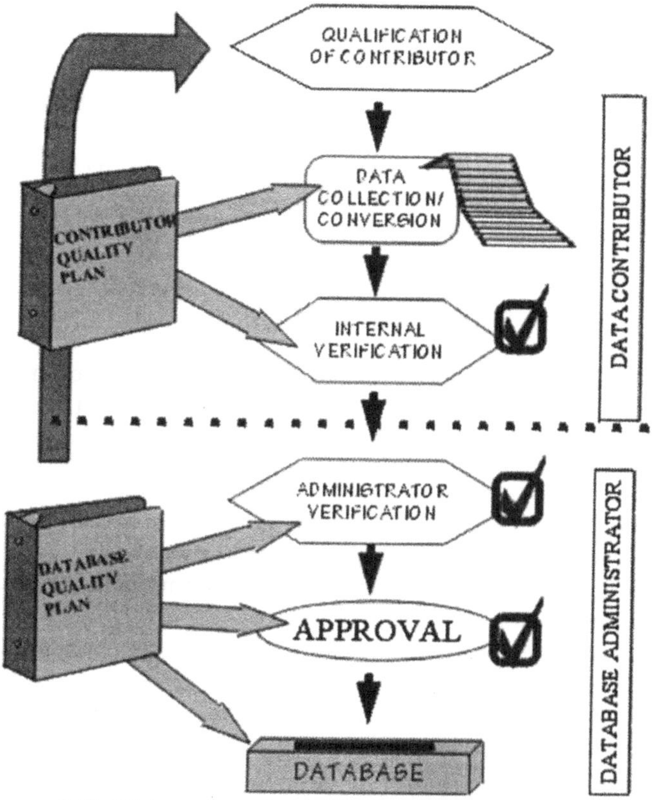

FIGURE 6.1. ***Overview of database quality system.***

6.5.1. Quality Plan for Database Administrator (DA)

The Database administrator (DA) will specify in detail the quality plan for the operation of the Database. This quality plan will provide details regarding the quality work procedures to ensure that the quality objectives are being met. The DA quality plan should include the following key elements:

- Definition of the Database mission
- Description of the scope of the Database
- Responsibilities of key individuals in the DA organization, with respect to the Database
- Basic policies for ensuring the quality of data
- Procedures for the handling of data subscriber's data, especially with regard to confidentiality of data

- Specification of data requirements
- Standardized reports for data analysis
- Procedure for verifying the quality of data prior to acceptance into the Database
- Procedures for reporting and handling of deviations
- Procedures for handling corrective actions
- Procedure for auditing data subscribers/data contributors

6.5.2. Quality Plan for Data Subscribers

Prior to submitting any data each data subscriber must submit a quality plan which meets the criteria established by the participant committee for review and approval. This plan should contain the following elements:

- Mission of quality plan
- Responsibilities of key individuals in the data subscriber organization, including any individual site data contributors
- The scope of the plan, including the equipment to be covered and the failure modes to be supported
- Work instructions/procedures for the acquisition/conversion of data
- System for reporting and processing of deficiencies
- Procedure for handling corrective actions
- Plan for complying with internal verification requirements
- Plan for maintaining certification, including approach to external audit verifications

6.5.3. Certification of Data Subscribers

Prior to any submission of data, a data subscriber must be certified as a supplier of quality data to the Database. The certification process entails the following steps:

- Review and approval of the Data Subscriber Quality Plan
- Review and approval of the collection/conversion system and procedures
- On-site inspection of data subscriber/data contributor records and data collection/conversion work process (see Appendix I, Section 1.3.2)
- Reporting and resolution of certification deficiencies

Once the data subscriber/data contributor has been certified, data can then be submitted to the Database administrator. At some time interval, the data subscriber must be recertified to confirm the continued quality of data submitted.

6.5.4. Internal Verification of Data Quality

Data subscribers/contributors will be expected to carry out internal verification of all data being submitted to the database. This verification process may differ from site-to-site; for example, the process may be manual, automated, or a combination of the two. Regardless, the process should contain these key elements:

- A list of items to check
- Acceptance criteria for above checklist

If data-conversion software is used to translate data from a maintenance database into the CCPS format, the correct version of that software must be verified and documented as part of the internal verification process.

A report summarizing the results of the internal verification check should accompany all data sets submitted to the CCPS Database administrator.

6.5.5. Verification of Data Prior to Acceptance

Once the data have been verified by the data subscriber/contributor, this information is then sent to the Database administrator for final quality verification. This will typically include a comprehensive check by a separate software package to reveal any missing required information, illegal codes, or data outside legal ranges. The sorts of checks to be performed will be specified in the DA quality plan. It is also essential that the checking software provides fixed criteria for acceptance of data.

As an example, Table 6.1 gives a listing of possible attributes to check, along with acceptance criteria. The objective of the checklist is to reveal as many obvious input errors as possible.

6.5.6. Recertification of Data Contributors

Recertification of data subscribers will be conducted at a minimum of every three years from the previous certification, and within three months of change of the data subscriber *at the discretion of the Database administrator*. Criteria for recertification are described in Appendix I, Section 1.3.3.

TABLE 6.1
Sample CheckList for Data Verification

Check Item	Acceptance Criteria
Check that all compulsory data fields are recorded	100% of required fields completed
Check that surveillance period begins after installation date	Minimum 1 day differential
Check that recorded failures occurred within reporting period	100% of failure data to occur within reporting period (no historical data)

6.5.7. Appeal Process

A facility may have a grievance with the administration of the Database. In most cases it is expected that problems will be resolved between the Database administrator and the facility. In extreme situations where the facility feels that the resolution to the issue is not satisfactory, the facility may appeal to a peer group, as described in the CCPS quality assurance procedure.

6.5.8. Audits of Work Process

From time-to-time, formal checks of the overall Data quality work process shall be performed. The purpose of these audits is to verify that the quality work process is fulfilling its objectives. These audits will be performed according to a set protocol. The initial audit will be performed by an independent third party.

Third-party audits will be conducted approximately once every two years. In the interim either the Database administrator or a third party will conduct at least one internal audit, and more if deemed necessary by the Database administrator to address any serious problems which may arise.

APPENDIX

I Guidelines for Data Collection and Submission

Authors' Note: *The contents of this appendix in particular can be expected to change with time. Therefore, this should be considered a "living" document, and this particular version a "snapshot" in time. While the overall Database structure and organization are not likely to change dramatically, details are likely to evolve. Possible future differences include:*

- ***Detail of data****—It is hoped and expected that ever-increasing levels of detail will be provided to the Database over time. This may result in additional text, and possibly a restructuring of the data to some extent.*
- ***Available analyses****—The Database participants may choose to fund additional software improvements (e.g., increased types of built-in statistical analyses), which would result in text changes.*
- ***Database administration****—It is very likely that the details of Database administration will change to reflect the actual vs. anticipated demands on the Database administrator, participant desires for increased/decreased administration support/oversight, external influences, etc.—that is, making the Database more efficient and effective.*

For these reasons and others, this is a working, not fixed, document. Current, updated knowledge on this appendix is available to Database Participants only.

I.1. Introduction

I.1.1. Data Types

The Database consists of three distinct data types, each of which may be linked to subgroups and from there to many data fields. The data types are subscriber, inventory, and event. These can be described as follows:

- *Subscriber*—This consists of information about the subscriber to the Database, including company name, contact names, etc.
- *Inventory*—This set of fields consists of several parts which identify the site, plants within that site, and its contents at the unit, system (i.e., equipment function), component (equipment item subsystem, e.g., lube oil system on a compressor) and part levels. In these fields, linkages to associated equipment are also made to allow data to be used for analyses of multiple systems that may be applicable.
- *Event*—These data describe specific activities that have taken place involving an equipment item, unit or for a plant as a whole. The event data will include equipment startups, voluntary shutdowns, failures, maintenance activities, and/or larger plant events which have shut down the equipment. A failure may or may not be associated with a maintenance activity ("corrective maintenance"), but an event may be recorded in the absence of a failure ("preventive maintenance," voluntary shutdown, power loss followed by restart, etc.). A subset of this data section will record specifics of loss of containment events, to allow cost analyses beyond those simply associated with equipment damage or downtime.

I.1.2. Subscriber Data

The Database will contain information regarding the data subscriber. These data will be used to document compliance of the subscribers with the data input requirements set forth in the database agreement. However the data, prior to distribution to other participating data subscribers, will be "anonymized" so that it will not be possible for subscribers to identify the source of the data, except from sites within the same organization. As a further safeguard, the Database will require that at least three companies are contributing to a selected system's data prior to the data being shared outside the company contributing the data.

I.1.3. Inventory Data

I.1.3.1. Characterization of Inventory Data

Equipment can be characterized as follows, similarly to ISO and other standards:

- *Identification data:* equipment location, classification, installation data, equipment unit data, etc.
- *Design data:* manufacturer's data, design characteristics, etc.
- *Application data:* operation, environment, etc.

In practice, the input tables are sorted through a hierarchy of equipment detail as well as the relationships between one system and related systems. An example of this is a "parent–child" relationship such as a compressor ("parent") having components including motor and a cooler ("children"). The information about equipment identification, design and application are provided and cross referenced at each level of equipment detail as appropriate.

I.1.3.2. Hierarchy of Inventory Data

Equipment inventory data are organized at two major levels of hierarchy:

- A user-defined "overview" level, which defines the general location and application of the equipment (the industry, site, plant and unit)
- A fixed level, which defines the equipment system and details of subordinate components, parts, and relationships to other equipment systems.

"Overview" Level

At a higher level, information may be gathered by industry, by site, by plant, or by unit. The distinction among these different levels is described below.

- *Industry*—The highest level of interest to the Database is "industry"; that is, those facilities producing similar product types. It is expected that users may wish to sort information at this level, to compare their performance to other comparable facilities.
- *Site/Plant*—A "site" refers to a plant or series of plants at a particular location. Often a "site" and a "plant" will be synonymous; for example, when there is a single, well-defined product line under one management structure. However, a site may consist of multiple plants. An example would be a major petrochemical complex. The distinction between a site and a plant is important to the user only, and so the distinctions between these two classes of hierarchy are defined

by the user in order to be consistent with internal data reporting needs, management structures, etc.

- *Unit Category*—A "process unit," for the Database purposes, refers to a collection of systems whose performance can be compared to similar sets of systems in other facilities. Examples include an ethylene cracker in an olefins plant, an alkylation unit in an oil refinery, or perhaps a tank farm. These process units can be characterized as one of the following unit "categories," for ease in organization across several types of industries:
  - Feed delivery
  - Feed treatment
  - Main process
  - Product treatment
  - Product delivery
  - Utility

Ultimately, the distinction of interest is that which allows different facilities to compare performance of sets of equipment which are conducting similar activities.

Fixed (System and Subsystem) Levels

In the fixed (equipment-level) hierarchy, the highest level is the "system," the second level is the "component", and the last is termed a "part." The terms are largely self-explanatory, but are described below:

- *System*—The highest level of equipment data entry is the System level. A "system" describes a specific function that a piece of equipment or intimately related group of equipment items is intended to perform to a limited number of material streams. Thus an ethylene cracking unit would not be a system, but a compressor and its associated utilities within the unit would be a system. Systems may be simple (e.g., a valve) or complex (e.g., compressor). At this time the following systems have been incorporated into the database, with work in progress on several others:
  - Compressor
  - Heat exchanger
  - Pressure relieving devices
- *Component*—The next level of indenture in the equipment system hierarchy is the Component level. A "component" is a subset of the system, and is necessary for the system to provide its function. It is not capable of providing the equipment function on its own. Thus a transmitter is a component of an instrument loop system.

- *Part*—The lowest level of indenture in the equipment system hierarchy is the Part level. A "part" may or may not be necessary for the component to operate properly; however its failure could result in the system progressing to a failure. In the example above, a capacitor would be a part within the component "transmitter." In pragmatic terms, a "part" can be thought of as the level at which an equipment item would be replaced if it failed.

I.1.3.3. System Boundaries

To define the components and parts of a system, it is necessary to define the system boundaries. For a given type of equipment, differences in opinion may exist on how this boundary should be drawn. Rules for this are provided in Appendix III, *Procedure for Developing System-Level Taxonomies*.

For the purposes of this Database, these boundaries are prespecified for each of the equipment types that are in the Database to assure consistency in treatment by the subscribers.

I.1.3.4. Components and Parts

The concepts of components and parts were defined earlier. They are best illustrated by way of an example, in this case the pressure relieving device system:

System Unit Subdivisions, RELIEF VALVE/RUPTURE DISK COMBINATION (partial list)

Components	Relief Valve	*Rupture Disc*	Instruments	Misc.
Parts	Disk	Disk	*Pressure gauge*	*Piping*
	Spring	Holder	*Pressure transmitter*	Pipe support
	Adjusting Screw			Flanges
	Stem			
	Bonnet			
	Body			
	Nozzle			
	Cap			
	Bellows			
	Vent			
	Seat			

Items in *italics* in the preceding table indicate items that can be considered systems as well ("subordinate systems").

I.1.3.5. Specific Inventory Tables

Following is a specific inventory example for one pressure relieving device case:

Conventional Spring Operated Relief Valve
Inventory Data Fields

Definition of Location and Inventory

Field Name	Description	Source
Subscriber ID	Unique name assigned to the entity supplying data to the database.	KEY FIELD
Subscriber Name		
Site ID (1)		
Site Name		
Plant ID (1)		
Plant Name		
Unit ID	Name of the unit operations associated with the system.	Pick List SUB4
Unit Category		
Unit Name		
System ID (1)	A unique identifier for the equipment item—most likely the equipment serial number.	User
System Grouping (1)	The name of the system type. The designated type links this system to the appropriate event and maintenance tables.	Pick List, SYS2 cascading to specific data sheet input forms
Equipment Tag Number (1)	The property or tag number assigned to the system by the subscriber.	User

Item Info Screen

Field Name	Description	Source
System ID (1)		Carry over from general input form
Relief Device Type (1)	Pressure Relief valve as generally specified in API RP520, figure G1, pressure relief valve specification sheet.	Cascading Pick List, SYS2. Carry over from general input form
Manufacturer (1)	Vendor	Pick list, REL3
Model Number		
Balanced bellows type		Yes and no check boxes
Auxiliary balanced piston	Only if balanced bellows type check box checked yes	Yes and no check boxes
Bonnet type		Pick list, REL8
Nozzle type		Pick list, REL9
Orifice designation	API letter orifice area for relief valves. Selection auto creates orifice area.	Pick list, REL4A
Orifice area	Actual area in square inches. Selection auto creates orifice designation.	Pick list, REL4B
Actual valve capacity		
Connection type	Type of inlet and outlet connection	Pick list, EQP2
Inlet size (1)	Nominal relief valve inlet diameter	User, inches or mm
Inlet connection pressure rating	Pressure rating (e.g., ANSI Flange Class)	User, EQP7
Outlet size	Nominal relief valve outlet diameter	User, inches or mm
Outlet connection pressure rating	Pressure rating (e.g., ANSI flange class)	User, EQP7

Application/Sizing 1 Screen

Field Name	Description	Source
Tag Number		Carry over from general input form
Set pressure	Relief set or opening or burst pressure	User, psig or barg
Accumulation		User, %
Cold set Pressure		User, psig or barg
Blowdown	Percent below set pressure that relief valve is expected to reseat	User, %
Fluid handled	Main fluid only	Pick List, SYS1
Fluid Phase at relief inlet	Phase of material to be relieved, at relief valve inlet at relieving conditions	Pick List, EQP3
Omega parameter	DIERS omega (two-phase or flashing liquids only)	User specified
Latent heat of vaporization		
Isentropic coefficient		
Inlet line pressure loss (design)	Percentage of set pressure	User, %
Specific gravity	Specific gravity of fluid at relief temperature	User specified
Molecular weight	Molecular weight of fluid handled	User specified
Compressibility factor	Compressibility factor (all vapor or gas fluids only)	User specified
Specific heat ratio	Ratio of specific heats (all vapor or gas fluids only)	User specified

Application/Sizing 2 Screen

Field Name	Description	Source
Tag Number		Carry over from general input form
Viscosity	Viscosity of fluid at operating temperature	User, centipoise
Design basis	Worst-case relief scenario for which the relief is designed	Pick List, REL7A (pressure) Pick List, REL7B (vacuum)
Code Category	Relief design standard, if applicable	Pick List, REL6
Allowable over pressure	Percent above set pressure	User, in %
Required capacity		
Required area		
Inlet pressure (normal)	Normal operating pressure at the inlet to the relief assembly	User, psig or barg
Back pressure built-up		User, psig or barg
Back pressure, constant superimposed	Back pressure at normal operating conditions.	User, psig or barg
Back pressure, variable superimposed	Back pressure at relieving conditions.	User, psig or barg
Operating temperature	Normal operating temperature at the inlet to the relief assembly	User, deg. F or deg. C
Flowing temperature	Temperature at the inlet of the relief assembly at relieving conditions	User, deg. F or deg. C
Fluid corrosiveness/ erosiveness	Qualitative assessment of the severity of the service	Pick List, EQP4
Fluid fouling potential	Qualitative assessment of the potential for the relief device to be stuck or plugged because of the material being handled in normal operation	Pick List, EQP5

Materials Screen

Field Name	Description	Source
System ID		Carry over from general input form
Body material	Material of construction of relief valve body	Pick list, EQP1A
Bonnet material	Material of construction of relief valve bonnet	Pick list, EQP1A
Spring material	Material of construction of relief valve spring	Pick list, EQP1A
Seat type		Pick list, REL5
Seat material	Material of construction of relief valve seat	Pick list, EQP1B
Guide material	Material of construction of relief valve guide	Pick list, EQP1A
Disk material	Material of construction of relief valve disc	Pick list, EQP1A
Adjusting Rings material	Material of construction of relief valve rings	Pick list, EQP1A
Washer material	Material of construction of relief valve washers	Pick list, EQP1A
Bellows material	Material of construction of relief valve bellows only if bellows check box checked yes	Pick list, EQP1A
Balanced piston material		Pick list, EQP1A
NACE 0175		Single check box

Accessories/Components/Misc Screen

Field Name	Description	Source
Cap type		Pick list, REL10
Lifting lever		Pick list, REL11
Gag		Single check box
Bug Screen		Single check box
Discharge treat-ment system	Treatment of relief effluent stream prior to release into atmosphere	Pick List , REF12
Piping support	Type of support used on relief assembly inlet/outlet piping	Pick List, EQP9
Protected by rupture disk?	If a rupture disk is entered as a system and recorded as a subordinate system, then the yes box should automatically be checked off	yes and no check boxes
Auxiliary Components	Multiple check boxes	REL13 (note 2)

Note 1: Bold italicized data fields are minimum required inventory information for 1998.

Note 2: Need ability to check off all auxiliary components that apply to installation

I.1.4. Event Data

I.1.4.1. Characteristics of Event Data

Events can consist of successes, failures and/or maintenance activities. These are characterized as follows:

Success Data:

- Identification of equipment
- Success data (e.g., correctly started up, or shut down, on demand)

Failure Data:

- Identification of failed equipment
- Failure data (e.g., failure date/time, failure mode, failure cause, parts failed, associated details for loss of containment events)

Maintenance Data:

- Identification of maintenance performed, with any linkage to a failure event ("corrective maintenance")
- Maintenance data (e.g., date, activity, parts maintained, man-hours, down time)

I.1.4.2. Linkage of Event Data

There will often, but not always, be linkages between failure events and maintenance events. There may also be a linkage between a failure and a loss of containment. The user may note a hierarchy of sorts between these categories. For example, upon an entry of a failure event, the user will be prompted as to whether there was an associated maintenance event or loss of containment event. The user should ensure that any interface software utilized by the site is capable of providing these linkages.

The event selection prompt has the following choices, depending on the type of system: (a) loss of containment, (b) startup, (c) shutdown, (d) curtailment, (e) inspection, (f) calibration, (g) maintenance [proof test, repair, preventive maintenance, calibration, condition-based maintenance], (h) process demand. An example of an event data table is provided in Section 2.1.3.

I.1.5. Data Analysis

Fundamentally, the goal of the Database is to provide the subscribers with information regarding the reliability and availability of plants and equipment in the various process industries. As information is added to the Database, the data will be numerous enough to allow progressively deeper sorting of equipment characteristics while maintaining statistical significance in the results.

Thus, after one year in the Database, it may be possible to obtain statistically significant results as to the frequency of compressor failures while running. The following year, it may be possible to sort by compressor type (e.g., centrifugal, reciprocating) at the same level of statistical significance. Ultimately, a subscriber should be able to sort by the characteristics of interest (e.g., compressor, centrifugal, hydrogen service, >2000 psig discharge pressure, manufacturer).

Some specific applications of this data analysis have been suggested, and include, among others,

- Determining reliability and availability of equipment, process units and plants
- Benchmarking one facility to another, within the same organization, or with industry as a whole
- Equipment selection
- Development of risk-based maintenance planning
- Equipment life-cycle cost

I.1.6. Database Limitations

When implementing and using a database that both obtains data from and provides data to a variety of users, it is necessary to develop database "structures" which assure that data are processed accurately and consistently. This inevitably results in some compromises. However, this Database has been designed to minimize the number of compromises in fundamental design. Participants may not use its full capability; for example, there are limits to the level of detail that can reasonably be expected to be entered into the Database given finite subscriber resources.

Equipment and data classifications will also not always be arranged the way a particular facility may treat them. For example, a compressor driver may be considered part of the compression system for one user, and may be considered an independent system in the eyes of another. Ideally, any differences between site and Database organization can be handled through site-developed data translation software.

Such philosophical differences are unavoidable, but to the extent possible attempts have been made to allow for them. In the case of input data limitations, provisions exist for the subscriber to enter data to a level of detail which their time permits—the return is in terms of the detail of information which can be output from the Database. Where equipment is arranged in a different manner of classification than a particular user would personally choose, provisions are made to allow linkages between equipment so that the information can be captured in the desired form.

Although considerable pains have been made to make this Database as useful as possible, the user can recognize that there will be continuing opportunities for improvement over the years to make it more useful and user-friendly. The Database administrator encourages comments on the effectiveness of this database, and suggestions for improvement. The contribution of time required on the part of the users to make this venture successful is appreciated, and whatever effort is applied to improving the Database will be recovered many times over in terms of subscriber time savings and technical benefits.

I.1.7. Goals of the Equipment Reliability Process

I.1.7.1. Usefulness of Data

The purpose of data collection is to try to gather information on significant design and operating variables that could affect the reliability of the equipment item of interest. When starting a database, it is desirable to allow for a large number of inputs, to make sure that nothing important is missed.

It is recognized that it is much easier to enable the collection of this information than it is to record it. For this reason the Database only *requires* a limited number of inventory fields, those that the participants feel can reasonably be obtained by all users. Similarly, the amount of event data required depends almost solely on the failure modes that the participant wishes to analyze—that is, the data required are those necessary to support the failure mode calculation of interest.

That said, it should be noted that this effort still goes beyond previous data collection efforts in terms of addressing operating variables. For example, in previous databases one might typically find that compressors are segregated by type (e.g., centrifugal, axial), perhaps power (e.g., horsepower), discharge pressure, or speed (rpm). This Database seeks to go further into variables in the operating environment, such as structural support, operating temperature, surrounding environment and more. These are known to be significant in their effects on reliability, yet in the past have, at best, been treated by an overall "environment" factor (severe, average, etc.).

I.1.7.2. Consistency of Data

One difficulty in developing a database is in defining equipment and operating variables in a manner that will be interpreted in the same way by all the users. For example, one facility may treat the motor on a compressor to be integral to the compressor, while another site may categorize them separately. For this reason, each system type is defined with a specific boundary indicated, which delineates which components are included and which are not.

More problematic may be the definition of process unit boundaries. As a simple example, an "alkylation unit" in an oil refinery may or may not include a prefractionation column, a butane isomerization process and other equipment, depending on the process requirements at the specific plant, or arbitrary administrative separations. In chemical plants, this problem can be even more acute.

Using the "unit category" feature allows relationships to be developed so that a more complex alkylation unit can be compared to a less complex one. Thus the Database provides a mechanism that allows participants to characterize plants in a consistent fashion.

I.1.7.3. Ease of Data Manipulation and Reporting

As noted earlier, a major concern during this Database development was to minimize the effort on the part of the user, to make sure that people will have enough time and energy to contribute data. A number of efforts are included in this Database to help achieve this end, including

- *Linked equipment*—where equipment is linked to other equipment, or where components are a subset of a system or are their own system, linkages are provided internally to the Database so that users do not have to enter information more than once. Thus, the fact that a compressor motor is located in an ethylene unit is extracted by the Database from the fact that the compressor is located in the ethylene unit, and that the motor is linked to the compressor.
- *Filtering*—The Database is designed to allow the user to sort for reliability or other outputs by any design or operating variable that has been input to the system. Thus there is no need to manually manipulate the data, or speculate on the percentage of failures which were of a certain type, etc.
- *Exporting*—The Database provides the ability to export filtered data to allow more advanced statistical analyses if the standard algorithms that are contained within the CCPS Database and described below are insufficient.
- *Built-in Analyses*—The Database provides some computational capabilities in addition to its primary task of data sorting and calculating "raw" equipment reliability information. These added tools are described in the next section.

I.1.7.4. Calculation Capabilities

The CCPS Database software provides some computational capabilities in addition to its primary task of supporting data quality assurance, data storage, filtering of data prior to analysis and export of this data for advanced statistical analysis. The algorithms contained in the software as of its first release are documented in the succeeding tables.

In general, it has been necessary to make simplifying assumptions. The major assumption is that failures are exponentially distributed, that is, exhibit constant failure rate. Failure modes representing continuous operations are analyzed as if the data were complete. In reality there will most likely be some amount of right censoring present. The assumption of complete data is made in this case in order to simplify the algorithm and is felt to be reasonable since the results will be somewhat conservative. Chapter 3 provides guidance on how the user can test the data to determine if the assumptions were indeed reasonable. Users who need more accurate analysis are encouraged to export the filtered data set and apply more rigorous statistical techniques.

Additionally, the software must determine which algorithm to use. This may change as a result of the specific failure mode being calculated or the normal operating mode of the equipment. This is the basis for making normal operating mode a required inventory data field. The software contains a rule set which automatically chooses the appropriate algorithm for a particular failure mode, once the data have been filtered.

Following are the calculations conducted by the CCPS software, and the data that are necessary to support the calculations:

Plant Level Analyses

Analysis Type	Fields Required to Support Analysis	Operating Mode	Algorithm to do Analysis
Average Shut-down Rate for Event Type *k* ($\lambda_{k,s}$)	• k • Ns • $ISED_i$ • $ISSD_i$	Continuous	$\lambda_{k,s} = \dfrac{Ns_k}{\sum_{i=1}^{n}(ISED_i - ISSD_i)}$
Mean Time to Shutdown for Event Type *k* ($MTTS_k$)	• k • $\lambda_{k,s}$	Continuous	$MTTS_k = 1/\lambda_{k,s}$
Average Failure Rate for Event Type *k* ($\lambda_{k,f}$)	• k • Nf • $ISED_i$ • $ISSD_i$	Continuous	$\lambda_{k,f} = \dfrac{Nf_k}{\sum_{i=1}^{n}(ISED_i - ISSD_i)}$
Mean Time to Failure for Event Type *k* ($MTTF_k$)	• k • $\lambda_{k,f}$	Continuous	$MTTF_k = 1/\lambda_{k,f}$
Mean Time to Restore for Event Type *k* ($MTTRs_k$)	• k • ED_i • Ns_k	Continuous	$MTTRs_k = \dfrac{\sum_{i=1}^{n}\sum_{j=1}^{Ns} ED_{ij,k}}{Ns_k}$
Availability for Event Type *k* (A_k)	• $MTTS_k$ • $MTTRs_k$	Continuous	$A_k = \dfrac{MTTS_k}{MTTS_k + MTTRs_k}$

Sample Analysis Table to Help Understand Algorithms

1. Failures per unit time for continuous operation of a plant, unit, system, etc.

System	#1	#2	#3	Total
Number of failures for a specific event type	5	0	2	7
Operating Time	5 yr	1 yr	3 yr	9 yr
$\lambda_{k,f}$				7 failures/9 years

Unit Level Analyses (same as Plant Level Analyses)

Analysis Type	Fields Required to Support Analysis	Operating Mode	Algorithm to do Analysis
Average Shut-down Rate for Event Type k ($\lambda_{k,s}$)	• k • Ns • $ISED_i$ • $ISSD_i$	Continuous	$\lambda_{k,s} = \frac{Ns_k}{\sum_{i=1}^{n}(ISED_i - ISSD_i)}$
Mean Time to Shutdown for Event Type k ($MTTS_k$)	• k • $\lambda_{k,s}$	Continuous	$MTTS_k = 1/\lambda_{k,s}$
Average Failure Rate for Event Type k ($\lambda_{k,f}$)	• k • Nf • $ISED_i$ • $ISSD_i$	Continuous	$\lambda_{k,f} = \frac{Nf_k}{\sum_{i=1}^{n}(ISED_i - ISSD_i)}$
Mean Time to Failure for Event Type k ($MTTF_k$)	• k • $\lambda_{k,f}$	Continuous	$MTTF_k = 1/\lambda_{k,f}$
Mean Time to Restore for Event Type k ($MTTRs_k$)	• k • ED_i • Ns_k	Continuous	$MTTRs_k = \frac{\sum_{i=1}^{n}\sum_{j=1}^{Ns} ED_{ij,k}}{Ns_k}$
Availability for Event Type k (A_k)	• $MTTS_k$ • $MTTRs_k$	Continuous	$A_k = \frac{MTTS_k}{MTTS_k + MTTRs_k}$

System Level Analyses

Analysis Type	Fields Required to Support Analysis	Operating Mode	Algorithm to do Analysis	Notes
Average Shut-down Rate for Event Type *k* ($\lambda_{k,s}$)	• k • Ns • ISED_i • ISSD_i	Continuous	$\lambda_{k,s} = \dfrac{Ns_k}{\sum_{i=1}^{n} (\text{ISED}_i - \text{ISSD}_i)}$	
	• k • Ns • ISED_i • ISSD_i • c_i	Cyclical	$\lambda_{k,s} = \dfrac{Ns_k}{\sum_{i=1}^{n} [(\text{ISED}_i - \text{ISSD}_i)c_i]}$	
Mean Time to Shutdown for Event Type *k* (MTTS_k)	• k • $\lambda_{k,s}$	Continuous, Cyclical	$\text{MTTS}_k = 1/\lambda_{k,s}$	
Average Failure Rate for Event Type *k* ($\lambda_{k,f}$)	• k • Nf • ISED_i • ISSD_i	Continuous	$\lambda_{k,f} = \dfrac{Nf_k}{\sum_{i=1}^{n} (\text{ISED}_i - \text{ISSD}_i)}$	The ISED is either the demand or the time that an item is removed from service for a proof test.
	• k • Nf • ISED_i • ISSD_i • c_i	Cyclical	$\lambda_{k,f} = \dfrac{Nf_k}{\sum_{i=1}^{n} [(\text{ISED}_i - \text{ISSD}_i)c_i]}$	
	• k • Nf • ISED_i • ISSD_i • c_i	Standby	$\lambda_{k,f} = \dfrac{Nf_k}{\left\{ \left[\sum_{i=f_1}^{n_f} (\text{ISED}_{i,f} - \text{ISSD}_{i,f})/2 \right] + \sum_{i=nf_1}^{n_{nf}} (\text{ISED}_{i,nf} - \text{ISSD}_{i,nf}) \right\}}$	
Probability of Failure to Start (P_{fts})	• k • Nf • $N_{st/d}$	Standby	$P_{\text{fts},k} = \dfrac{Nf_k}{N_{st/d}}$	

Analysis Type	Fields Required to Support Analysis	Operating Mode	Algorithm to do Analysis	Notes
Mean Time to Failure for Event Type k ($MTTF_k$)	• k • $\lambda_{k,f}$	Continuous	$MTTF_k = 1/\lambda_{k,f}$	
Mean Time to Restore for Event Type k ($MTTRs_k$)	• k • ED_i • Ns_k	Continuous	$MTTRs_k = \frac{\sum_{i=1}^{n}\sum_{j=1}^{Ns} ED_{ij,k}}{Ns_k}$	Based on Tag No.
Mean Time to Repair, Active for Failure Type k ($MTTRa_k$)	• ART_i	Continuous, Cyclical, Standby	$MTTRa_k = \frac{\sum_{i=f_1}^{nf} ART_i}{Ns_k}$	Based on Tag No.
Mean Time to Repair, Total (MTTRt)	• TRT_i	Continuous, Cyclical, Standby	$MTTRt_k = \frac{\sum_{i=f_1}^{nf} TRT_i}{Ns_k}$	Based on Tag No.
Availability for Event Type k (A_k)	• $MTTS_k$ • $MTTRs_k$	Continuous	$A_k = \frac{MTTS_k}{MTTS_k + MTTRs_k}$	

Sample Analysis Table to Help Understand Algorithms

2. Failures per cycle for cyclic operation of an equipment system, etc.

System	#1	#2	#3	Total
Nf	2	4	1	7
Surveillance Interval	2 yr	2 yr	1 yr	
Number of cycles per year	10,000	12,000	8,000	
Product of Surveillance Interval andNumber of cycles per year	20,000	24,000	8,000	52,000
$\lambda_{k,s}$				7 failures/52,000 cycles

Component Level Analyses (same as System Level Analyses)

Analysis Type	Fields Required to Support Analysis	Operating Mode	Algorithm to do Analysis	Notes
Average Shutdown Rate for Event Type *k* ($\lambda_{k,s}$)	• k • Ns • $ISED_i$ • $ISSD_i$	Continuous	$\lambda_{k,s} = \dfrac{Ns_k}{\sum_{i=1}^{n}(ISED_i - ISSD_i)}$	
	• k • Ns • $ISED_i$ • $ISSD_i$ • c_i	Cyclical	$\lambda_{k,s} = \dfrac{Ns_k}{\sum_{i=1}^{n}[(ISED_i - ISSD_i)c_i]}$	
Mean Time to Shutdown for Event Type *k* ($MTTS_k$)	• k • $\lambda_{k,s}$	Continuous, Cyclical	$MTTS_k = 1/\lambda_{k,s}$	
Average Failure Rate for Event Type *k* ($\lambda_{k,f}$)	• k • Nf • $ISED_i$ • $ISSD_i$	Continuous	$\lambda_{k,f} = \dfrac{Nf_k}{\sum_{i=1}^{n}(ISED_i - ISSD_i)}$	The ISED is either the demand or the time that an item is removed from service for a proof test.
	• k • Nf • $ISED_i$ • $ISSD_i$ • c_i	Cyclical	$\lambda_{k,f} = \dfrac{Nf_k}{\sum_{i=1}^{n}[(ISED_i - ISSD_i)c_i]}$	
	• k • Nf • $ISED_i$ • $ISSD_i$ • c_i	Standby	$\lambda_{k,f} = \dfrac{Nf_k}{\left\{\left[\sum_{i=f_1}^{n_f}(ISED_{i,f} - ISSD_{i,f})/2\right] + \sum_{i=nf_1}^{n_{nf}}(ISED_{i,nf} - ISSD_{i,nf})\right\}}$	
Probability of Failure to Start (P_{fts})	• k • Nf • $N_{st/d}$	Standby	$P_{fts,k} = \dfrac{Nf_k}{N_{st/d}}$	

Analysis Type	Fields Required to Support Analysis	Operating Mode	Algorithm to do Analysis	Notes
Mean Time to Failure for Event Type k ($MTTF_k$)	• k • $\lambda_{k,f}$	Continuous	$MTTF_k = 1/\lambda_{k,f}$	
Mean Time to Restore for Event Type k ($MTTRs_k$)	• k • ED_i • Ns_k	Continuous	$MTTRs_k = \frac{\sum_{i=1}^{n}\sum_{j=1}^{Ns} ED_{ij,k}}{Ns_k}$	Based on Tag No.
Mean Time to Repair, Active for Failure Type k ($MTTRa_k$)	• ART_i	Continuous, Cyclical, Standby	$MTTRa_k = \frac{\sum_{i=f_1}^{nf} ART_i}{Ns_k}$	Based on Tag No.
Mean Time to Repair, Total (MTTRt)	• TRT_i	Continuous, Cyclical, Standby	$MTTR_k = \frac{\sum_{i=f_1}^{nf} TRT_i}{Ns_k}$	Based on Tag No.
Availability for Event Type k (A_k)	• $MTTS_k$ • $MTTRs_k$	Continuous	$A_k = \frac{MTTS_k}{MTTS_k + MTTRs_k}$	

Part Level Analyses

Analysis Type	Fields Required to Support Analysis	Operating Mode	Algorithm to do Analysis	Notes
Average Shut-down Rate for Event Type ***k*** ($\lambda_{k,s}$)	• k • Ns • $ISED_i$ • $ISSD_i$	Continuous	$\lambda_{k,s} = \dfrac{Ns_k}{\sum_{i=1}^{n}(ISED_i - ISSD_i)}$	
	• k • Ns • $ISED_i$ • $ISSD_i$ • c_i	Cyclical	$\lambda_{k,s} = \dfrac{Ns_k}{\sum_{i=1}^{n}[(ISED_i - ISSD_i)c_i]}$	
Mean Time to Shutdown for Event Type ***k*** ($MTTS_k$)	• k • $\lambda_{k,s}$	Continuous, Cyclical	$MTTS_k = 1/\lambda_{k,s}$	
Average Failure Rate for Failure Type ***k*** ($\lambda_{k,f}$)	• k • Nf • $ISED_i$ • $ISSD_i$	Continuous	$\lambda_{k,f} = \dfrac{Nf_k}{\sum_{i=1}^{n}(ISED_i - ISSD_i)}$	
	• k • Nf • $ISED_i$ • $ISSD_i$ • c_i	Cyclical	$\lambda_{k,f} = \dfrac{Nf_k}{\sum_{i=1}^{n}[(ISED_i - ISSD_i)c_i]}$	
	• k • Nf • $ISED_i$ • $ISSD_i$ • c_i	Standby	$\lambda_{k,f} = \dfrac{Nf_k}{\left\{\left[\sum_{i=f_1}^{n_f}(ISED_{i,f} - ISSD_{i,f})/2\right] + \sum_{i=nf_1}^{n_{nf}}(ISED_{i,nf} - ISSD_{i,nf})\right\}}$	
Mean Time to Failure for Event Type ***k*** ($MTTF_k$)	• k • $\lambda_{k,f}$	Continuous	$MTTF_k = 1/\lambda_{k,f}$	

Analysis Type	Fields Required to Support Analysis	Operating Mode	Algorithm to do Analysis	Notes
Mean Time to Restore for Event Type k ($MTTRs_k$)	• k • ED_i • Ns_k	Continuous, cyclical, standby	$MTTRs_k = \frac{\sum_{i=1}^{n}\sum_{j=1}^{Ns} ED_{ij,k}}{Ns_k}$	Based on Tag #
Availability for Event Type k (A_k)	• $MTTS_k$ • $MTTRs_k$	Continuous, cyclical, standby	$A_k = \frac{MTTS_k}{MTTS_k + MTTRs_k}$	

I.1.8. Certification of a Subscriber

I.1.8.1. Company Certification

Prior to certifying a subscriber, the company must first be a participant. Company participation is a simple process of signing the Database agreement and satisfying the financial obligations.

An individual subscriber is brought into the system by applying for certification. The certification process includes the following steps:

1. Preregistration—
   (a) The applicant provides the following:
   - Name of facility (e.g., "DEPCO New Brunswick Specialty Chemicals")
   - Address
   - Contact person/phone
   - Description of the facility
     - General description of processes/products (about 1–3 pages)
     - Industry type (e.g., specialty chemical, oil refining)

   (b) the Database administrator provides the following:
   - Data subscriber confidential ID
2. Verification—
   (a) The Database administrator verifies that the company is a participant in good standing
3. Complete application—
   (a) The applicant provides the following:
   - Desired date of entry to the Database (generally, the application should be made about two months in advance of desired entry)
   - Data Subscriber Quality Plan

This is a progression of steps taking place in chronological order, as depicted below.

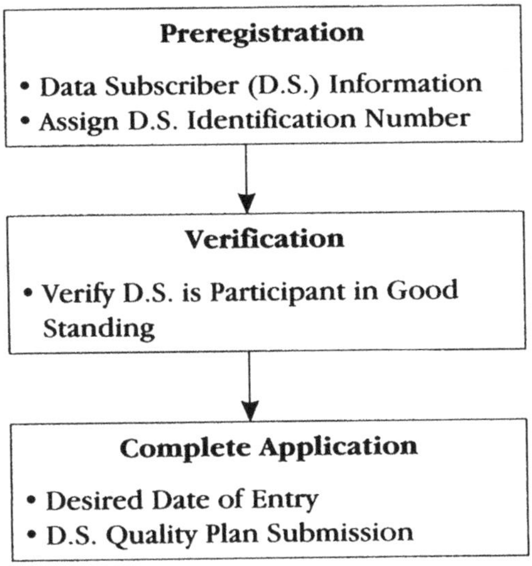

I.1.8.2. Site Visit(s)

Prior to initial submission of data, the Database administrator will review the data subscriber's proposed quality plan to assure that it will meet the Database criteria. When satisfied, the Database administrator will either:

(a) visit the site (or sample site(s), if the data subscriber is responsible for submitting data for a number of sites) and inspect records and corresponding proposed data submissions to verify that the information is being collected in a complete, consistent and accurate manner, or

(b) accomplish the same by having the data subscriber submit raw data sources that can be cross-checked against proposed data submissions.

The verification in Option (a) is preferred for ***initial*** certification, and is as follows:

- Visit with the data subscriber, to verify that goals and requirements of the Database are understood.
- Observe the site's raw record keeping and verification of examples of any subordinate-provided data.
- Observe the data provider entering data into the Database.

It is recognized that Option (b) will be more economical for data subscribers that are remotely located relative to the Database administrator. In the case of Option (b), however, the burden of proof is on the data subscriber to demonstrate to the satisfaction of the Database administrator a level of quality equivalent to that provided by Option (a).

The Database administrator will provide any corrective suggestions and guidance necessary to assure compliance. Following approval of the Database administrator, the site's data will be accepted by the Database.

A (re)visit to a site may be made in some situations, including the following:

- A large number of nonconformities have been detected
- There is a change in the site data contributor or data subscriber, without enough time for transfer of knowledge from the prior data contributor/subscriber

Typically, if either of the above have occurred, the site will be suspended from both data submission and information retrieval from the Database until such time as the Database administrator is satisfied that the replacement data subscriber/data contributor is functioning adequately. Visits to sites may be made under different circumstances as those described above; only if data integrity is questioned will a site be suspended from the data contribution work process.

I.1.8.3. Certification Updates

On occasion, each data subscriber and its associated quality plan will be recertified. Recertification may be prompted by the following factors:

- Change in quality plan
- A high level of nonconformities over a sustained period of time

In general, the preferred approach for recertification will be that described in (b) in Section 1.3.2.

Acceptance will be determined by use of the certification procedure. Philosophically, the Database administrator will be required to ensure the integrity of the Database, which includes both monetary and technical aspects, for the mutual benefit of the participants. If a particular company or site performs in a way that compromises the integrity of the Database, this would be grounds for denial of recertification. An example might be a situation in which a site has so many nonconformities that the Database is compromised, or that it is an undue financial burden on the Database (i.e., the other participants) to maintain the integrity of the Database in the presence of the nonconformities.

I.2. Data Definitions and Descriptions

I.2.1. Introduction

I.2.1.1. Taxonomy

A "taxonomy" is the structure by which data and systems are organized in a database. Chapter 5 and Section I.1.3 of this appendix describe the taxonomy used in this Database. The taxonomy includes elements of data hierarchy (or "indenture").

Again, in the Database it will be possible to sort information at each level of the hierarchy. It will also be possible to obtain information based on several operating variables, specific to each type of system. These variables are recorded in the inventory descriptions (see Section 2.2.2).

I.2.1.2. Boundaries

System boundaries are predefined in the Database to ensure that users do not "compare apples to oranges." To repeat an earlier example, one plant may consider the motor on a large compressor to be part of the compressor "equipment," whereas another plant may not. An arbitrary decision on including the driver or not could certainly impact resulting reliability data. For this reason the Database is reasonably strict in defining precisely what constitutes a particular "system." Appendix III, Guidelines for Developing System Taxonomies, describes the approach for setting equipment boundaries.

The boundaries for each type of system in the Database are described in Section 2.2. Please note that these classifications must be rigorously adhered to in order for the database to be accurate!

I.2.2. Inventory Descriptions and Data Sheets

The following sections describe some of the systems that have been defined and imbedded within the Database. Sample data sheets are provided for illustrative purposes. However, the Database utilizes drop-down lists and relational "views" (nested windows) to query the user for information, rather than require the user to fill out the following sheets. Complete copies of all current data sheets are maintained by the Database administrator, and distributed to the Database participants as updates occur.

I.2.2.1. Compressors

Inline Centrifugal Compressor Inventory Data Fields

Note: Inventory Data fields for inline centrifugal compressors based upon API617

Screen:
Title: In-Line Centrifugal Compressor

Screen Header Info		
Field Name	**Description**	**Source**
Subscriber Name		
Site Name		
Plant Name		

System Information Tab

Field Name	**Description**	**Source**
Equipment Tag (1)	The property or tag number assigned to the system by the subscriber.	User
Unit Name		
System Type (1)	= Inline Centrifugal Compressor	Final choice of cascading pick list SYS2. See description
SystemID (1)		
Normal Operating Mode (1)		Pick List SYS3
Maintenance Strategy		Pick List SYS7

Operational Information Tab

Field Name	**Description**	**Source**
Date Installed		
Surveillance Start Date (3)		
Subordinate to other system	Not applicable (?)	check box
Associated system	Only if subordinate checked yes—SEE ABOVE	
Operating Time	Calculated field	
Calendar Time	Calculated field	

Application Information 1 Tab

Field Name	Description	Source
Fluid handled (1)	Main process gas only	Pick List, SYS1A
Molecular weight	Molecular weight of gas handled	User supplied, in g/mole
Specific gravity	Specific gravity of gas at relief temperature	User specified
Relative humidity	Relative humidity of the gas	User supplied, in %
Fluid corrosiveness	Qualitative assessment of the severity of the service.	Pick List, EQP4A
Fluid erosiveness		Pick List, EQP4B
Gas flow rate (1)	Total gas flow rate	User supplied, in SCFH
Inlet pressure, design (1)	Pressure at the inlet to the first stage of the compressor	User supplied, in psia
Inlet temperature, design	Temperature at the suction of the first stage	User supplied, in deg. F
Inlet $C_p/C_v(k)$	Ratio of specific heats at the compressor inlet	
Inlet compressibility (z)	Compressibility of the gas at the compressor inlet	

Application Information 2 Tab

Field Name	Description	Source
Sidestream flow?	Bring up sidestream table if yes	Single check box for yes
Discharge pressure, design (1)	Pressure at the outlet of the last stage of the compressor	User supplied, in psia
Discharge temperature, design	Temperature at the outlet of the last stage	User supplied, in deg. F
Discharge $C_p/C_v(k)$	Ratio of specific heats at the compressor	
Discharge compressibility (z)	Compressibility of the gas at the compressor	
Input power (1)	Shaft input power to compressor	User supplied, in bhp
Speed	Compressor rotative speed	User supplied, in rpm
Polytropic efficiency		User supplied, in %
Hazardous electrical classification?		Single check box for yes

If Side Stream Flow checked 'yes'

Sidestream pressure		User, in psia
Flow		User, in SCFH
Flow direction	Check boxes for 'in' or 'out'	

Control Strategies Tab

Field Name	Description	Source
Suction Throttling	Provide linkage to subordinate systems if yes checked	Single check box for yes
Inlet Guide Vane	Provide linkage to subordinate systems if yes checked	Single check box for yes
Variable Speed	Provide linkage to subordinate systems if yes checked	Single check box for yes
Discharge Blowoff	Provide linkage to subordinate systems if yes checked	Single check box for yes
Cooled Bypass	Provide linkage to subordinate systems if yes checked	Single check box for yes
Anti-surge	Provide linkage to subordinate systems if entry made	Pick list, COM21

Construction Information Tab

Field Name	Description	Source
Manufacturer (1)	Vendor	Pick list, COM1
Model Number		
Applicable standard	Compressor design standard(s), if applicable	Pick List, COM2
Driver type	Provide linkage to subordinate systems	Pick List, COM3
Transmission type	Power transmission type	Pick List, COM4
Number of impellers	Dictates number of records in the next associated table	

Component Information Begins Here

Individual Impeller Design Sub Tab(s)
(One set of records for each impeller)

Field Name	Description	Source
Diameter		
Number of vanes per impeller		
Impeller type		Pick List, COM19
Fabrication type		Pick List, COM20
Impeller material		Pick list, EQP1A
Maximum Yield Strength		User supplied, in psi
Maximum hardness		User supplied, in BHN
Minimum hardness		User supplied, in BHN
Tip internal width		User supplied, in inches
Maximum Mach Number @ Eye		User supplied
Maximum Impeller Head @ 100% Speed		User supplied, in inches

Casing Design Tab

Field Name	Description	Source
Casing split type		Pick List, COM17
Casing material		Pick list, EQP1A
Thickness		User supplied, in inches
Corrosion allowance		User supplied, in inches
Maximum working pressure		User supplied, in psig
Maximum design pressure		User supplied, in psig
Test pressure		User supplied, in psig
Test fluid		Pick List, COM18
Maximum operating temperature		User supplied, in deg. F
Diaphragm material		Pick list, EQP1A

Rotor Design Tab

Shaft Info

Field Name	Description	Source
Shaft material		Pick list, EQP1A
Diameter @ Impellers		User supplied, in inches
Diameter @ Coupling		User supplied, in inches
Maximum Yield Strength		User supplied, in psi
Hardness		User supplied, in BHN
Maximum Torque Capability		User supplied, in ft-lbs

Balance Piston Info

Balance piston material		Pick list, EQP1A
Area		User supplied, in in^2
Fixation method		Pick List, COM22

Shaft Sleeves Info if checked yes

@ Interstage close clearance points—material		Pick list EQP1A
@ Shaft seals—material		Pick list EQP1A

Internal Labyrinths Design Tab

Field Name	Description	Source
Interstage type		Pick list COM23
Interstage material		Pick list EQP1A
Balance piston type		Pick list COM23
Balance piston material		Pick list EQP1A

Shaft Seal Design Tab

Field Name	Description	Source
Type		Pick list COM7
Supplementary device type		User supplied
Settling out pressure		User supplied, in psig
Buffer gas system used?	Bring up buffer gas system table if yes	Single check box for yes

If Buffer Gas System checked 'yes'

Field Name	Description	Source
Type of buffer gas		Pick List, SYS1B
Maximum buffer gas flow		User supplied, in lbs/min
Buffer gas pressure		User supplied, in psig

Bearing Housing Tab

Field Name	Description	Source
Type		Pick list COM24
Split?		Y or N
Material		Pick list EQP1A

Radial Bearing Design Tab

Field Name	Description	Source
Bearing span		User supplied, in inches

Inlet End Info

Type		Pick list COM8
Manufacturer		Pick list COM25
Length		User supplied, in inches
Shaft Diameter		User supplied, in inches
Unit Load—actual		User supplied, in psi
Unit Load—allowable		User supplied, in psi
Base Material		Pick list EQP1A
Babbitt Thickness		User supplied, in inches
Number of Pads	Only applicable to type = tilting pad	User supplied
Load Orientation	Only applicable to type = tilting pad	Pick list COM26
Pivot Location	Only applicable to type = tilting pad	Pick list COM27

Exhaust End Info

Field Name	Description	Source
Type		Pick list COM8
Manufacturer		Pick list COM25
Length		User supplied, in inches
Shaft Diameter		User supplied, in inches
Unit Load—actual		User supplied, in psi
Unit Load—allowable		User supplied, in psi
Base Material		Pick list EQP1A
Babbitt Thickness		User supplied, in inches
Number of Pads	Only applicable to type = tilting pad	User supplied
Load Orientation	Only applicable to type = tilting pad	Pick list COM26
Pivot Location	Only applicable to type = tilting pad	Pick list COM27

Thrust Bearing Design Tab

Active Side Info

Field Name	Description	Source
Type		Pick list COM9
Manufacturer		Pick list COM25
Unit Load—actual		User supplied, in psi
Unit Load—allowable		User supplied, in psi
Area		User supplied, in in^2
Number of Pads	Only applicable to type = tilting pad	User supplied
Pivot Location	Only applicable to type = tilting pad	Pick list COM27
Pad Base Material	Only applicable to type = tilting pad	Pick list EQP1A
Lubrication		Pick List COM28
Thrust Collar Type		Pick List COM29

Inactive Side Info

Field Name	Description	Source
Type		Pick list COM9
Manufacturer		Pick list COM25
Unit Load—actual		User supplied, in psi
Unit Load—allowable		User supplied, in psi
Area		User supplied, in in^2
Number of Pads	Only applicable to type = tilting pad	User supplied
Pivot Location	Only applicable to type = tilting pad	Pick list COM27
Pad Base Material	Only applicable to type = tilting pad	Pick list EQP1A
Lubrication		Pick List COM28
Thrust Collar Type		Pick List COM29

Instrumentation Tab

Field Name	Description	Source
Bearing Temperature Devices	Provide linkage to subordinate systems if yes checked	Single check box for yes
Radial Shaft Vibration Devices	Provide linkage to subordinate systems if yes checked	Single check box for yes
Axial Shaft Position Devices	Provide linkage to subordinate systems if yes checked	Single check box for yes
Casing Vibration Devices	Provide linkage to subordinate systems if yes checked	Single check box for yes

End of Component Information

Planned Maintenance Tasks (Note 4)—FUTURE POTENTIAL SCOPE

Field Name	Description	Source
Task Number		
Task Type		
Task Name		
Scheduled hours		

Notes:

1. Bold italicized data fields are minimum required inventory information for 1998.
2. Tabs containing required field information should be colored yellow.
3. Surveillance start date is required for calculations to proceed, but should not be required for inventory records to be saved as part of documenting inventory. Possibly show it as a different color.
4. Both yes and no check boxes are provided except for repair which only has a yes box. Neither box checked represents lack of information in the Database, that is, unknown whether a particular strategy is employed.

I.2.2.2. Pressure Relieving Devices

[See example in Section I.1.3.5]

I.2.2.3. Heat Exchangers

Shell and Tube Heat Exchanger Inventory Data Fields

Definition of Location and Inventory

Field Name	Description	Source
Subscriber ID	Unique name assigned to the entity supplying data to the database	KEY FIELD
Subscriber Name		
Site ID (1)		
Site Name		
Plant ID (1)		
Plant Name		
Unit ID	Name of the unit operations associated with the system	Pick List SUB4
Unit Category		
Unit Name		
System ID (1)	A unique identifier for the equipment item—most likely the equipment serial number	User
System Grouping (1)	The name of the system type. The designated type links this system to the appropriate event and maintenance tables	Pick List, SYS2 cascading to specific data sheet input forms
Equipment Tag Number (1)	The property or tag number assigned to the system by the subscriber.	User

Item Info Screen

Field Name	Description	Source
System ID (1)		Carry over from general input form
Manufacturer (1)	Vendor	Pick List, HX3A
Model Number		
Design pressure (shell)		User, psig or bar
Design pressure (tube)		User, psig or bar
Design temperature (shell)		User, deg. F or C
Design temperature (tube)		User, deg. F or C
Number of passes per shell		User
Corrosion allowance		User, inches or mm
Connection size:		
Shell in		User, inches or cm
Shell out		User, inches or cm
Tube in		User, inches or cm
Tube out		User, inches or cm
Intermediate in		User, inches or cm
Intermediate out		User, inches or cm
Tubes:		
Number of tubes		User
Length of tubes		User, feet or meters
Tube thickness (ave.)		User, inches or mm
Outer diameter		User, inches or mm
Pitch dimension		User, inches or mm
Pitch type		Pick List, HX7
Tube type		Pick List, HX9
Tubesheet support		User, pick either "Stationary" or "Floating"
U-bend?		Yes/No

Field Name	Description	Source
Shell:		
Outer diameter		User, inches or mm
Thickness		User, inches or mm
Channel or bonnet?		User, pick "Channel" or "Bonnet"
Floating head cover?		Yes/No
Baffles:		
Cross type?		Pick List, HX9
Long—seal type?		Pick List, HX10

Application/Sizing 1 Screen

Field Name	Description	Source
Tag Number		Carry over from general input form
Fluid handled, hot side	Main fluid only. "Hot side" refers to the side of the exchanger which is hotter at the inlet.	Pick List, SYS1
Fluid handled, cold side	Main fluid only. "Cold side" refers to the side of the exchanger which is colder at the inlet.	Pick List, SYS1
Fluid rate, hot side	Total fluid flow rate	User supplied, in pounds per hour
Fluid rate, cold side	Total fluid flow rate	User supplied, in pounds per hour
Phase at inlet, hot side	Phase of material at the inlet to the exchanger. "Hot side" refers to the side of the exchanger which is hotter at the inlet.	Pick List, EQP3
Phase at outlet, hot side	Phase of material at the outlet of the exchanger. "Hot side" refers to the side of the exchanger which is hotter at the inlet.	Pick List, EQP3
Phase at inlet, cold side	Phase of material at the inlet to the exchanger. "Cold side" refers to the side of the exchanger which is colder at the inlet.	Pick List, EQP3
Phase at outlet, cold side	Phase of material at the outlet of the exchanger. "Cold side" refers to the side of the exchanger which is colder at the inlet.	Pick List, EQP3
Molecular weight, hot side	Molecular weight of fluid handled. "Hot side" refers to the side of the exchanger which is hotter at the inlet.	User supplied, in g/mole.

Field Name	Description	Source
Molecular weight, cold side	Molecular weight of fluid handled. "Cold side" refers to the side of the exchanger which is colder at the inlet.	User supplied, in g/mole.
Fluid velocity, hot side	Nominal velocity of fluid through the exchanger	User supplied, in feet per second
Fluid velocity, cold side	Nominal velocity of fluid through the exchanger	User supplied, in feet per second
Inlet pressure, hot side	Pressure at the inlet to the hot side of the exchanger.	User, in psig
Inlet pressure, cold side	Pressure at the inlet to the cold side of the exchanger.	User, in psig

Application/Sizing 2 Screen

Field Name	Description	Source
Tag Number		Carry over from general input form
Allowable pressure drop, hot side	Pressure drop available on the hot side of the exchanger.	User supplied, in psi
Allowable pressure drop, cold side	Pressure drop available on the cold side of the exchanger.	User supplied, in psi
Calculated pressure drop, hot side	Calculated pressure drop on the hot side of the exchanger.	User supplied, in psi
Calculated pressure drop, cold side	Calculated pressure drop on the cold side of the exchanger.	User supplied, in psi
Fluid corrosiveness/ erosiveness	Qualitative assessment of the severity of the service.	Pick List, EQP4
Corrosion allowance, hot side	Extra thickness allotted to the hot side of the exchanger beyond that required to meet design pressure.	User supplied, in inches
Corrosion allowance, cold side	Extra thickness allotted to the cold side of the exchanger beyond that required to meet design pressure.	User supplied, in inches
Fluid fouling factor, hot side	Qualitative assessment of the potential for the hot side of the exchanger to be severely fouled or plugged because of the material being handled in normal operation.	Pick List, EQP5
Fluid fouling factor, cold side	Qualitative assessment of the potential for the cold side of the exchanger to be severely fouled or plugged because of the material being handled in normal operation.	Pick List, EQP5

Field Name	Description	Source
Heat exchanged (design)	Amount of heat exchanged at design conditions.	User supplied, in BTU/hour
Inlet temperature (design), hot side	Temperature at the inlet of the hot side of the exchanger.	User, in deg. F
Outlet temperature (design), hot side	Temperature at the outlet of the 'hot side' of the exchanger. "Hot side" refers to the stream which is hotter at the inlet to the exchanger.	User, in deg. F
Inlet temperature (design), cold side	Temperature at the inlet of the cold side of the exchanger.	User, in deg. F
Outlet temperature (design), cold side	Temperature at the outlet of the 'cold side' of the exchanger. "Cold side" refers to the stream which is colder at the inlet to the exchanger.	User, in deg. F
Category	Heat exchanger design standard, if applicable.	Pick List, HX6

Materials Screen

Field Name	Description	Source
System ID		Carry over from general input form
Shell material	Material of construction, shell	Pick List, EQP1A
Tube material	Material of construction, tube	Pick List, EQP1A
Gasket material, shell side	Material of construction, shell gasket	Pick List, EQP1A
Gasket material, head	Material of construction, head gasket	Pick List, EQP1A
TEMA class		Pick List, HX11A/B/C

Accessories/Components/Misc Screen

Field Name	Description	Source
Auxiliary Components	Multiple check boxes	HX12 (note 2)

Note 1: Bold italicized data fields are minimum required inventory information for 1998.

Note 2: Need ability to check off all auxiliary components that apply to installation.

I.2.3. Event Descriptions and Data Sheets

"Events" are described in Section I.1.4, but basically can be considered to be either "voluntary" or "involuntary." A voluntary event will typically be an equipment startup, an unscheduled or scheduled test, scheduled maintenance, or voluntary shutdown for process reasons. ***However, if a failure requiring a work order or equivalent is detected during scheduled maintenance, the maintenance will be recorded as a failure.*** An example is a running test of a spare pump, which discovers that the spare is not available as expected. An involuntary event results from some sort of failure, and is recorded as such.

Examples of the event information requested for a pressure relieving device are shown in the following tables. Actual event data will be provided when the data fields are agreed upon for a particular item. In practice the user will be guided through the information request process through prompts and pick lists in the software.

Loss of Containment Input Forms—Relief Device

System Information Tab

Field Name	**Description**	**Source**
Subscriber Name (1)		
Site Name (1)		
Plant Name (1)		
Unit Name		
System ID (1)		
Equipment Tag (1)	The property or tag number assigned to the system by the subscriber.	User
Event Start Date		
Event Start Time		
System Type		Auto fill from system ID
Applicable Event		
Description		

Loss of Containment Tab

Field	Description	Source
Event end date	Actual time relief reseats due to return to controlled pressure.	
Event Duration or Event end time (Depending what is picked, the other should be calculated)		User
Pressure at time of failure	Maximum pressure attained within pressure circuit.	Actual pressure (psig) or unknown
Fluid Released		Pick List, SYS1
Severity		Pick list, REF4
Quantity Released, lbs		
Hole Size	Input hole diameter if known directly and calculate area field.	
Estimated Area	Input estimated area and calculate equivalent hole size diameter.	
Event description	Memo field	
Consequence		Pick List, PLF7.1
Reportable Incident?		(Yes/No)
If reportable, reportable to:		Pick List, PLF8
Comments		User
RESULTS		
Failure Mode		Determined by Rule Set
FAILURE CAUSES		
Failure Cause(s)	Causes need to be linked with failure modes that show up in the results section.	Pick List REF3

Inspection Tab

Field	Description	Source
Discharge Piping visual inspection	Note some relief valves are tied into a header system where the discharge line is inaccessible for internal inspection.	Pick list, REF2
Weep hole visual inspection		Pick list, REF2
Evidence of seat leakage		Check box for both yes and no options
External leakage		Check box for both yes and no options
Repair required	Need link to repair input form and applicable components or parts. (One yes check box is all that is required. Unchecked is no by default.)	Yes/no
Pressure between rupture disk and relief valve	APPLICABLE WHEN RUPTURE DISK AND RELIEF DEVICE USED IN TANDEM.	psig
Insulation loss?	Only if insulation is identified as part of this relief system.	Pick list, REF5
Heat tracing operating?	Only if heat tracing is identified as part of this relief system.	Check box for both yes and no options
Pressure monitor functional?	Only if monitor is identified as part of this relief system.	Check box for both yes and no options
Temperature monitor functional?	Only if monitor is identified as part of this relief system.	Check box for both yes and no options
Excess flow check valve functional?	Only if excess flow check valve is identified as part of this relief system.	Check box for both yes and no options
Comments	Memo field	User
RESULTS		
Performance Results		Determined by Rule Set
FAILURE CAUSES		
Failure Cause(s)	Causes need to be linked with failure modes that show up in the results section.	Pick List REF3

Demand Tab

Field	Description	Source
Event end date	Actual time equipment pressure returns to normal.	
Event Duration or Event end time (Depending what is picked, the other should be calculated)		User
Reason for process demand	If equipment failure, then need link to equipment system events.	Pick List, REL7A
Event description	Memo field	
Valve lift? (Discharge through the relief valve)	Relief to normal discharge location. Automatically "Yes" if lift pressure entered.	Check box for 'Yes' and 'No' options. If 'Yes', then link to additional controlled discharge input fields below
Lift Pressure		User, psig or barg or unknown
Maximum process pressure achieved	1. This field should be automatically filled if Lift pressure is recorded. (May not be able to do automatically because pressure maximum equipment pressure, lift pressure, and relieving pressure may be different.) 2. Required field if NO checked in Valve lift field.	User, psig or barg and allow option for text entry (e.g., >100 psig or unknown) to account for cases where upper limits in pressure transmitters are exceeded or not recorded
Rapid cycling of the valve (chatter) audible during incident?		Check box for both yes and no options
Valve reseat?		Check box for both yes and no options
Reseat Pressure		User, psig or barg
Code deviation? [FUTURE}	Was the maximum allowable pressure of protected equipment allowed by code exceeded?	Check box for 'Yes' or 'No'. (Automatically a 'Yes' if Maximum Equipment pressure >1.5 * equipment design pressure or >1.5 * relief valve set pressure. This probably needs to be modified in the future)

Field	Description	Source
Fluid Relieved		Pick List, SYS1A
Quantity Relieved, lbs		
Discharge treatment successful?		Yes or no If no, then text field to describe failure (need to expand this in future)
Consequence		Pick List, PLF7.1
Reportable Incident?		(Yes/No)
If reportable, reportable to:		Pick List, PLF8
Protected equipment damaged?	Provide link to other equipment systems.	Check box for 'Yes' or 'No'. (automatic 'Yes' if Loss of Containment = 'Yes').
Loss of containment?	Release to atmosphere other than normal relief pathway. Link to loss of containment input form.	yes/no (one yes check box is all that is required. Unchecked is no by default)
Repair or replacement required?	Link to repair input form if "Yes" checked.	yes/no (one yes check box is all that is required. Unchecked is no by default)
Proof Test performed	Link to proof test input form if "Yes" is checked.	yes/no (one yes check box is all that is required. Unchecked is no by default)
Comments	Any other information deemed appropriate.	User
RESULTS		
Leak Ratio	ratio = leak pressure/set pressure.	Calculated if applicable
Lift Ratio	ratio = lift pressure/set pressure.	Calculated
Reseat Ratio	ratio = reseat pressure/set pressure.	Calculated
Performance Results		Determined by Rule Set
FAILURE CAUSES		
Failure Cause(s)	Causes need to be linked with failure modes that show up in the results section.	Pick List REF3

Proof Test Tab

Field	Description	Source
Proof Test Date	Actual from general event input form.	
Event Time	Actual from general event input form. Don't really need this for proof tests. Could simply use a default value..	
Visual Flow Path Inspection: Inlet		Pick List REF2
Visual Flow Path Inspection: Outlet		Pick List REF2
External leakage	If yes, need linkage to part events.	Need both a yes and a no check box
Seat Leakage	Automatically 'Yes' if leak pressure less than 95% of set pressure.	Need both a yes and a no check box
Leak Pressure		Actual pressure (psig)
Valve lift?	Automatically yes if lift pressure entered.	Check box for both yes and no options
Lift Pressure		User, psig or barg
Maximum test pressure	1. This field should be automatically filled if Lift pressure is recorded. 2. Required field if NO is checked in Valve lift field.	User, psig or barg
Valve reseat		Check box for both yes and no options
Reseat Pressure	If yes checked off for "Valve Reseat."	User, psig or barg
Leakage following reseat		Need both a yes and a no check box
Maintenance Performed?	Link to repair input form.	
Actual test time (hours)		
Additional information	Any other information deemed appropriate.	

Field	Description	Source
RESULTS		
Leak Ratio	Ratio = leak pressure/set pressure	Calculated if applicable
Lift Ratio	Ratio = lift pressure/set pressure	Calculated
Reseat Ratio	Ratio = reseat pressure/set pressure	Calculated
Proof Test Results/Failure Mode(s) if applicable		Determined by Rule Set
FAILURE CAUSES		
Failure Cause(s)	Causes need to be linked with failure modes that show up in the results section.	Pick List REF3

Repair Tab

Field	Description	Source
Maintenance Start Date	Actual from general event input form.	
Maintenance Start Time	Actual from general event input form. Don't really need this for proof tests. Could simply use a default value.	
Date maintenance completed		User
Maintenance hours expended	Actual hours performing repair work. Does not include time to order parts or to restore service, etc.	User
Maintenance performed	This needs to be a check off all that apply.	Pick list, REM1
Components repaired	Only if component repair chosen as one of maintenance activities.	Pick list, REL13
Components replaced	Only if component replaced chosen as one of maintenance activities. Check off all that apply.	Pick list, REL13
Part replaced	Only if part replaced chosen as one of maintenance activities. Check off all that apply.	Pick list, REL14

Field	Description	Source
Component failure mode	FUTURE	Determined by failure mode rule set (Future)
Component Failure Cause(s)	FUTURE	
Part failure mode	FUTURE	Determined by failure mode rule set (Future)
Part Failure Cause(s)	FUTURE	
Additional information	Any other information deemed appropriate.	
Disposition following repair	Only applicable if not scrapped.	Pick list, REM2

If disposition following maintenance is return to service then these fields need to be filled in:

Date Returned to Service	Date: Actual date accepted for service.	
Time Returned to Service	Time: Actual Time accepted for service.	

Note that the above tables are examples only. Some fields and tables will change, depending on the function of the equipment involved.

I.3. Data Flow

Data flowing into the Database undergoes a series of transformations and checks to ensure that it meets the Database requirements. The data, once aggregated, is available for retroactive correction if needed. For quality-verified data, the data is anonymized, distributed to data contributors and available for search requests by participants. A flow path for data is depicted in Figure I.3.1.

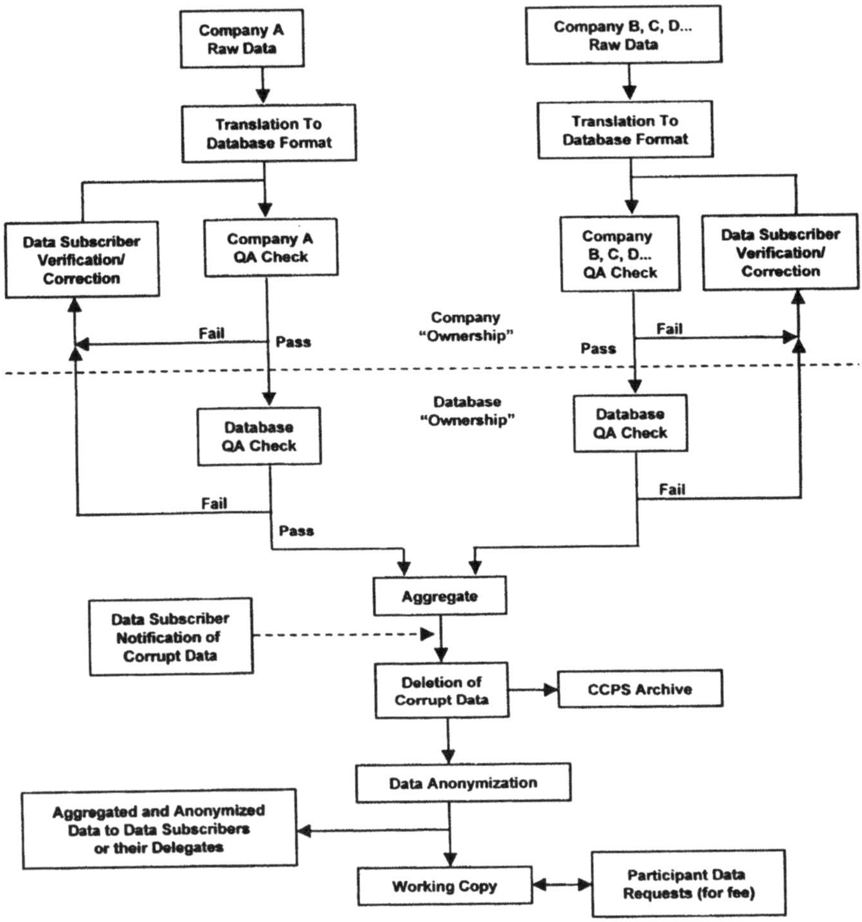

FIGURE I.3.1. *Data flow.*

I.4. Internal Quality Assurance of Data

I.4.1. Data Input Organization

The planning process encompasses two phases of the Database operation: startup and ongoing. Each of these is discussed below. Note that the quality standards are the same for each phase, and are assured by application of the data subscriber's quality plan in conjunction with the Database administrator quality plan.

Initial Input Phase

In the startup phase, the data subscribers/data contributors will be examining existing records to determine if historical documentation of equipment performance meets the minimum standards of the Database. For example, is it known with a high level of confidence when and for how long every shutdown event took place?

Naturally, written (or electronic) records are of paramount importance in establishing an equipment history. It is recognized that there will be some oral history of failures that were not recorded, maintenance which was undoubtedly conducted as part of a schedule but that was not documented, etc. In general, it is expected that such oral data will not meet the quality standards of the Database, and so should not be used as a primary data source. It is conceivable that an oral history may be used to correct nonconformities in written/electronic data, however.

Ongoing Phase

Before embarking on the Database effort, a site must recognize that significant initial resources will need to be devoted to the effort to establish site software development and integration which will ultimately allow the data processing effort to be minimal. Some pains have been made to minimize the up front effort, without compromising the usefulness of the data. The current balance of these conflicting goals still results in a noticeable initial workload, and the site should take the opportunity of discussing with other users their approaches to minimizing these efforts, and observe their data collection in advance to gain a realistic understanding of the time commitment needed. As time passes, this effort should drop off and ultimately be limited to nominal amounts of time in performing self-checks and responding to external requests, as discussed elsewhere.

I.4.2. Quality Plan

Each data subscriber will be required to submit a quality plan that must meet the minimum requirements described in the Quality Plan for Data Subscribers document. The quality plan will describe, among other topics, data quality assurance through use of training, self-checks, and reporting of nonconformities. These topics are briefly described below.

I.4.2.1. Training

Participant personnel who are responsible for providing data are also responsible for obtaining the knowledge necessary to ensure that the data meets the needs of the Database. At minimum this will consist of having read the entirety of the relevant documentation prepared during the Database development effort. Ideally, this also consists of formal orientation training provided by a CCPS-approved training supplier.

Different levels of training may be relevant for different people involved in the Database effort. A company's designated data subscriber should understand the history and philosophy of the Database as well as the workings of the software. A data contributor at a site, on the other hand, may only need to know how to collect the raw data and ensure that it is properly transferred into the Database format.

I.4.2.2. Self-Checks

Self-checks are the first point of assuring quality data. Performing quality self-checks will not only result in good quality data, but will also minimize Database oversight and costs. It will also ultimately result in lower costs for the user, in not having to respond to nonconformity checks, costs associated with remedial site visits from the Database administrator, etc. The required self checks are described below.

I.4.2.2.1. Before Data Processing Begins

Document Orientation

The user should be familiar with all Database documentation, in order to gain the proper perspective of project goals, data needs, use of data, etc. This will give an appreciation of the basis for the Database and its guidelines, and the rigor needed in following the guidelines and other instructions.

Systems Identification and Familiarization

For the systems to be included in the Database, the user should verify that the site's system meets the criteria set forth by the Database, especially in regards to the boundaries around the system. These boundaries are not necessarily the same as the plant might use internally. Each system should be provided a unique identifier (see data input sample sheets, Section I.1.3.5). This will facilitate electronic data mapping.

The data contributor/subscriber should also be knowledgeable in the equipment systems for which data is to be provided. This does not necessarily mean the provider needs to be the "expert" on a piece of equipment, but the person should be able to access and interpret design, installation and performance information, and should have access to the "expert."

Software Orientation

The data subscriber should have tested the data input software to understand how to gain access and maneuver in the program. The data subscriber/data contributor will ideally understand the meaning of the prompts given by the program, although this is not mandatory since in an electronic system the critical issue is performing a proper translation of the site's database information in the CCPS format. The Database administrator or designee will demonstrate and test this ability during the initial site certification process, or as a result of training certification.

Subordinate Duties

It is expected that the data subscriber may depend on others to gather some of the information. The data subscriber should verify that these data contributors have the information needed, and time commitment and necessary training to perform their functions.

Problems and Deviations

The data subscriber's quality plan should indicate the direction to turn if there is a problem. The ultimate resource will be the Database administrator, but it should be recognized that these services will not always be instantly available, and so it is useful to develop alternate resources such as users at other sites or companies.

All deviations from normal practice should be communicated to the Database administrator. This communication may be formal or informal as the circumstances may suggest, although the Database administrator may require formal documentation for the Database's records. Any "significant" database problems, even if resolved by an alternate source, should also be copied to the Database administrator. This will help verify that the problem was solved properly, and will provide useful information to the Database administrator of problems in the system. Such feedback can also be posted to the other data subscribers for the benefit of all.

I.4.2.2.2. During Data Collection

Data Collection Timeliness and Accuracy

The data subscriber should verify that data are being collected in a prompt and accurate manner. This includes verification that system boundaries are being "maintained"; that is, that information is being provided for the system within the boundaries and not also for "services" outside the boundaries.

Typographical Errors

While some checks will be provided inside and outside of the software, it is of paramount importance that the data provider check their entries for typographical errors, since there are no screens which can detect some types of errors.

Follow-up Check
The user may wish to check the accuracy of data input by extracting the same information out of the Database. The Database should be provided some period of time (~30 days?) to assure that the data have been received and processed.

I.4.3. Deviation Reporting

The following are among the deviations which should be reported to the Database administrator:

- *Expected delays in data submission*—There are good reasons why a facility might fall behind in providing data. Among these might be an extended emergency leave, data provider leaving the employment of the site, etc. If this is expected to jeopardize the requirements for supplying data, this should be communicated with the Database administrator as soon as possible.
- Discovered errors in data which have been supplied
- Problems in operating the database

A standard form for deviation reporting is provided in the Quality Plan for Data Subscribers.

I.5. Data Submission Procedure

I.5.1. User Software Requirements

Each data contributor will be supplied with software that will allow data to be entered either manually via pull-down menus or electronically via site management systems. This software will be designed to operate with a common electronic software tool.

I.5.2. Data Submission Protocol

The Database will accept data at all times, although updates to the Database will be made quarterly based on data received by March 31, June 30, September 30, and December 31 of each calendar year. Data contributors may submit data quarterly, semiannually or annually, as long as each years' minimum data requirements are satisfied.

I.5.3. Format of Data Submission

Since it is expected that the participants employ a variety of management information systems, the data from each site will be collected by site-supplied software which will convert the fields into a format compatible with the CCPS Database. The user can submit this file either by email or by sending a disk to the Database administrator. Once the data have been received and verified as acceptable for use, the Database administrator will confirm with the site the accepted receipt of the data.

I.5.4. Availability of New Data

The Database administrator will be responsible for updating the CCPS Database quarterly, with updates expected about one month following the end of the previous data submission deadline. The Database will provide a date on the Database entry page which will indicate the date of the most recent update.

I.6. External Quality Assurance of Data

I.6.1. Field Verification

As part of the certification process described earlier the Database administrator will typically visit the site (or a sample site, in the case where a single Data subscriber is collecting data from multiple sites) to assure that the site will provide quality data. See Section 1.3.2 for details.

I.6.2. Spot Checks

On occasion, the Database administrator may choose to "observe" a specific block of data which has been submitted by a site. These data may be analyzed to determine if they appear to be accurate (that is, no typographical errors). These will be conducted for all sites at an interval described in the Quality Plan for Database Administrator, and may be conducted more frequently for any sites having a history of deviations

I.6.3. Receipt and Logging

The Database administrator receives and logs incoming data as follows:

- Identification of the site, and equipment for which data is being provided, is recorded in an electronic "ledger" that notes the following:

the date of data entry, the equipment for which data are provided, the time frames covered by the data (e.g., is the start date for these data the same as the end date for the previous data submission), and the site providing the information. The ledger contains no technical information, and is simply used to verify that the site is meeting the terms of the agreement for providing data.

- Technical data are extracted from the data provided. An anonymous company "tag" is assigned to the data. This allows the Database to extract any data that may subsequently be found to be corrupted. In all other respects the data are "anonymized." Provisions will be made for a company to repopulate the anonymized database with its own, more detailed, location information upon receipt of the fully anonymized database from the CCPS Database administrator.

I.6.4. Validation

Validation of data that have been received will take several forms, including:

- *Data Completeness*—Any required data which is missing from any entry will prompt a deviation and will send a message back to the data contributor flagging the missing entry
- *Illegal Data*—Any entries which are not permitted (such as text entry where a numerical entry is required, or where event data is provided which is not covered in the quality plan) will be identified in the software and flagged to the data contributor for verification or correction.
- *"Reasonableness" check*—Any data which is suspicious (e.g., operating temperature 5000 degrees) will be returned to the Data Contributor for verification.

Again, there is nothing the Database can do to detect errors within the realm of reasonableness (e.g., operating pressure of 30 psig instead of 300 psig). This is the reason it is a good idea for the data subscriber to conduct a self-check by extracting out the data that has been provided, as described earlier.

I.6.5. Procedures for Nonconformities

Data inputs that appear to be faulty are handled as described above on a case-by-case basis. Any submissions that have the potential to compromise the integrity of the Database results will be classified as nonconformities and documented by the Database administrator. If there is a repeated history of nonconformities, the site will be warned that the level of nonconformities is reaching an unacceptable level.

Note: The phrase "a repeated history of nonconformities" is relative and may be changed periodically to assure that each Database participant is treated in an equitable manner. It is anticipated that a new Data subscriber may have a relatively large number of nonconformities relative to the amount of data being submitted—the Database administrator will be looking for evidence of "continuous improvement" for data subscribers who have a disproportionate number of nonconformities.

At some point of repeated nonconformance, the effort required on the part of the Database administrator to assure the integrity of the Database will be greater than the value of the data provided. At this point the Database administrator will develop qualitative or quantitative measures by which to determine when to invoke the following procedures.

If the negative performance continues, this may prompt additional oversight by the Database administrator, including requests for detailed support documentation, verification of site quality plan effectiveness, additional site visits, additional spot checks, etc. Any additional costs borne by the Database in performing such extraordinary oversight will be charged to the responsible site as a condition for staying in the Database. If this additional oversight is not sufficient to ensure Database integrity, or if the additional time demands on the Database administrator compromise the administrator's other responsibilities to the Database, this will be considered grounds for the termination of the site's certification.

I.7. Data Contribution and Data Distribution

It is an operating principle of the CCPS Database that the data contributors will be allowed to contribute information at a level which best suits their resources and information needs, at or above a minimum level. Distribution of data will occur as per data contributor agreements or exception rules established by the data contributors. These issues are described in more detail below.

I.7.1. Data Contribution

For the 1998 calendar year, each of the data contributors has agreed to provide one or more of the following:

- Data for 8 operating years on compressors, or
- Data for 50 operating years on heat exchangers, or
- Data for 100 operating years on relief devices

Alternatively, a data contributor may provide a total of 100 years operating experience on any combination of the types of equipment above. In addition, a data contributor must be able to support two failure modes for the each of the equipment for which data is provided.

I.7.2. Data Distribution

I.7.2.1. Data Distribution to Data Contributors

The aggregated, anonymized data will be distributed as a populated CCPS database software tool according to the following rules:

- Any participant who meets minimum requirements for data submission gets all raw anonymized data.
- In future years, minimum requirements for data submission are expected to increase.
- In the future, when the minimum requirements are increased, new participants, and participants unable to meet the minimum data submission requirements, will be able to retrieve a level of data that is commensurate with what is provided, according to a formula that will be developed when necessary.

I.7.2.2. Requesting Data from the CCPS Database Administrator

There are cases where participants are unable to contribute data (e.g. E&C firm), or in which data contributors want to retrieve data beyond the level at which they have provided data. In these cases, the participants will be required to submit a Data Search Request Form (with purchase order or other payment method) to the Database administrator. The Database administrator will perform the search and send the results to the participant.

APPENDIX

II Sample Pick Lists for Data Fields

This appendix presents a sample of the pick lists used in data fields throughout the Equipment Reliability Database. It is intended to provide a sample of the data populating some of the standard fields. This appendix is not meant to provide a complete listing of all data available in all pick lists.

II.1. Pick Lists Related to Upper Level Taxonomy

External Environment
Indoors—controlled
Indoors—not controlled
Other
Outdoors—controlled environment
Outdoors—enclosed
Outdoors—not enclosed
Outdoors—partially enclosed
Submerged in Liquid
Underground—controlled
Underground—not controlled
Unknown

Agency
National Government Agency
State or Province Agency
Local Agency

Industry Name
Chemicals
Electronics
Food
Industrial Gas
Oil/gas
Pharmaceuticals
Power Generation
Pulp & Paper
Steel Production

II.2. Pick Lists Related to Plant Inventory and Events

Operating Mode
Batch
Continuous Operation
Cyclical Operation
Standby

Shutdown Reasons
Cooling Water Curtailment
Customer Request
Electrical Curtailment
Interlock Response to Plant Upset
Loss of Cooling Water Utility
Loss of Plant Electrical Feed
Loss of Steam Utility
Planned Maintenance
Spurious Interlock Shutdown
Steam Supply Curtailment

Failure Description
Fitting Failure
Gasket Blowout
Gasket Leak
Packing Blowout
Packing Leak
Pipe Leakage
Pipe Rupture
Seal Blowout
Seal Leak
Vessel Leakage
Vessel Rupture

Event Types
Curtailments
Inspections
Loss of containment
Maintenance
Process upsets
Shutdowns

Curtailment Reason
Cooling Water Curtailment
Customer Request
Electrical Curtailment
Equipment Failure
Maintenance
Steam Supply Curtailment
Weather Conditions

Loss of Containment Severity
Major leak
Major spill/release
Minor leak
Minor spill/release
Rupture

Consequence	Subtype
Safe Discharge	N/A
Explosion	BLEVE
Explosion	Vapor Overpressure
Fire	Pool Fire
Fire	Jet Fire
Fire	Oxygen Fire
Fire	Fire Ball
Fire	Flash Fire
Fire	Metal Fire
Fire	Other Fire
Explosion	Runaway Polymerization
Explosion	Decomposition
Explosion	Vapor Cloud
Explosion	Internal deflagration or
Explosion	Other (specify)
Other (specify)	N/A
Potential toxic exposure	N/A
Potential Asphyxiation	N/A
Equipment damage	N/A
Hygiene issue	N/A
Environmental issue	N/A
Excessive noise	N/A

II.3. Pick Lists Related to General Equipment Information

Metal Material of Construction		
Aluminum	Monel	Nickel plated carbon steel
Brass	Hastelloy C	Stainless steel
Bronze	High temperature alloy	Titanium
Carbon Steel	Inconel X750	Alloy steel
Cast Iron	Iron	Chrome Moly Steel
3.5 % Nickel Steel		

Phase
Gas or vapor
Liquid (nonflashing)
Liquid (subcooled)
Two-phase (gas or vapor/liquid)
Two-phase (solid/liquid)
Two-phase (gas or vapor/solid)
Two-phase (immiscible liquids)
Three-phase (gas or vapor/liquid/solid)

Pressure Class Rating
125 # ANSI
150 # ANSI
250 # ANSI
300 # ANSI
600 # ANSI
900 # ANSI
1500 # ANSI
2500 # ANSI

Equipment Class ID	Equipment Class
1	Control and Instrumentation
2	Electrical
3	Fire Fighting
4	Fired Equipment
5	Heat Exchange
6	Piping
7	Pressure Relieving Device
8	Rotating Machinery
9	Separation Equipment
10	Solids Handling
11	Utilities
12	Valves
13	Vessels

Fluid Corrosiveness
Clean service
Moderate service
Severe service

Fouling Potential
Normal environment at relief inlet and/or outlet is corrosive
Normal environment at relief inlet and/or outlet can autopolymerize
Potential for solids to buildup
Material can freeze when cooled to ambient temperatures
Negligible
Other

II.4. Pick Lists Related to Relief Devices

Consequence	
Controlled safe discharge	Fire—fireball
Explosion (other, specify)	Fire—flash fire
Explosion—BLEVE	Fire—jet fire
Explosion—decomposition	Fire—metal fire
Explosion—internal deflagration or detonation	Fire—other fire
Explosion—runaway polymerization	Fire—oxygen fire
Explosion—vapor cloud explosion	Fire—pool fire
Explosion—vapor overpressure	Other (specify)
	Uncontrolled Discharge

Cap Type
Screwed
Bolted

Lifting Lever Type
Plain
Packed
None

Discharge Treatment
Tail pipe to atmosphere
Flare
Thermal oxidizer
Other thermal/ combustion unit
Scrubber
Catch tank
Other (specify)

Auxiliary Components
Insulation
Steam heat trace
Electric heat trace
Temperature monitor
Pressure monitor
Barrier Device

Visual Inspection Result
Unobstructed—Clean
Partially obstructed
Plugged
Unobstructed—Dirty
Not Inspected

Failure Cause
Failed/worn component or part
Inadequate Design
Manufacturer error
Mishandled
Misused

Relief Valve Parts		
Bellows	Guide seal	O-ring retainer
Bellows gasket	Guide/cover	O-ring seat seal
Blowdown ring	Inlet flange	Outlet flange
Body	Jam nut screw	Packing ring
Body gasket	Lift stop	Spring
Bonnet	Lift stop ring	Spring adjusting screw
Bonnet gasket	Lock wire	Stem
Bug vent housing	Lock screw (blowdown ring)	Stem insert
Cap	Lock screw (disk holder)	Stem retainer
Cap gasket	Lock screw gasket	Stud
Disc	Nozzle	Support plate
Disc holder	Nozzle seal	Test lever
Drain plug	Nuts	Test lever fork
Guide bushing	O-ring	Other (specify)

Relief Valve: Spring-Loaded Subtype
Conventional safety relief valve
Balanced bellows safety relief valve
Balanced bellows safety relief valve with auxiliary balanced piston

Relief Device Seat Type
Metal to metal
Soft(elastomeric)
Other

Bonnet Type
Open
Closed

Nozzle Type
Full
Semi

Insulation Inspection Results
Partially missing/ loss
Missing/ complete loss

Severity
External leakage
Seat leakage
Spuriously Opens
Equipment Rupture

Manufacturer		
GEC Alsthom Sapag	Fitting Valve and Control Corp.	Rexarc
Anderson Greenwood (Agco)	Fluid Mechanics Valve Co.	Rockwood Sevenderman Corp.
Aquatrol	Frick Co.	Richards
Axelson	Fukui Seisakusho Co.	Sarasin Industrie
IMI Bailey Birkett	Fulflo	Seetru
Baird Manufacturing	Hansen Technologies Corp.	Cyrus Shank Co.
Beacon Brass Co.	Henry Valve Co.	Sherwood
Bestobell	Hydro-Seal Valve Co.	Steuby Manufacturing Co.
Birkett	ITT Bell and Gossett	Superior Valve Co.
Bopp & Reuther	Jayco	Target Rock Corp.
BPS	Keystone Valve Ltd.	Taylor Tools
Campfield Hausfeld	F.C. Kingston Co.	Texsteam Inc.
IMI Cash Valve Inc.	Kunkle	Tomco Division
Cavagna Group S.P.A.	Ledco	Triangle-Sempell
Circle Seal Controls	Lesser	Trist
Conbraco Industries	Lonergan	Valvtechnologies
Con. Pneumatic	Mercer Valve Co	Watts Fluidair
Consolidated	Mogas Industries	Division of Watts Regulator Co.
Control Devices Inc.	Motoyama Engineering Works	Watts Industries
Crosby	Mueller Brass Co.	Wellmark Corp.
Dresser	Pignone	Whessoa
Eaton Corp.	R.S.B.D.	Wilkins Regulator Co.
Econosto	Refrigerating Specialties Division	Other
Farris (Teledyne Fluid Eng.)/Fisher	Rego (Engineered Controls International Inc.)	

Orifice Size ID	Orifice Size Square Inches	Orifice Size ID	Orifice Size Square Inches
4	0.049	K	1.838
6	0.11	L	2.853
8	0.196	M	3.6
D	0.11	N	4.34
E	0.196	Other	0
F	0.307	P	6.38
G	0.503	Q	11.05
H	0.785	R	16
J	1.287	T	26

Applicable Design Code
ANSI/ASHRAE 15
API RP-520/521
API Std. 2000
ASME Section I (Power boilers)
ASME Section VIII (Pressure Vessels)
ASME/ANSI B31.1(Power piping)
ASME/ANSI B31.2(Fuel gas piping)
ASME/ANSI B31.3(Process piping)
ASME/ANSI B31.4(Liquid hydrocarbon, LPG, ammonia, alcohols piping)
ASME/ANSI B31.5(Refrigeration piping)
CGA S1.1
CGA S1.2
CGA S1.3
NFPA 30
None
Other (specify)

Overpressure Design Basis
Blocked outlet, continued flow into protected equipment
Loss of cooling or reflux
Accidental mixing (entrance of volatile material into hot location)
Overfilling
Abnormal vapor input
Accumulation of noncondensable
Abnormal heat input
Internal explosion
Pressure surge
Chemical reaction or decomposition
Thermal expansion
Fire—external
Failed heat exchanger tube
Control failure
Other (specify)

Maintenance Activities Performed
Adjusted blowdown
Adjusted set pressure
Cleaned
Disassemble
Lapped
Machine seats
None
Repair component
Replace component
Replace part
Retest
Scrapped

Disposition Following Maintenance
Returned to service
Stored in inventory
Scrapped

II.5. Tables Related to Compressors

Curtailment Reasons
Compressor failure
Customer request
Force of nature
Human error
Insufficient demand
Planned maintenance
Process upset
Utility failure

Shutdown Reasons
Compressor failure
Customer request
Force of nature
Human error
Insufficient demand
Planned maintenance
Process upset
Utility failure

Failure to Start Reasons
External utility unavailable
Human error
Process part failure
Subsystem failure

Sub Reasons
Compressor subsystems failed
Compressor parts failed

Reason(s) for Maintenance
Curtailed throughput
High bearing temperature
High discharge pressure
High discharge temperature
High noise level
High oil temperature
High vibration
Leakage
Low discharge pressure
Low oil level
Low oil pressure

Action Performed
Adjustment/ calibration
Change oil
General lubrication
None
Overhaul
Repair
Replace component
Replace parts
Replace system
Scrapped and not replaced
Scrapped and replaced

Disposition Following Repair
Returned as an improved condition
Returned to as new condition
Temporary fix with degraded performance
Temporary fix with full performance capability

Compressor Failure Descriptors
High bearing temperature
High discharge pressure
High discharge temperature
High motor amps
High oil temperature
High suction pressure
High vibration
Low discharge pressure
Low oil level
Low oil pressure
Low suction pressure
Process related shutdown

Compressor Process Pressure-Containing Parts
Casing
Cooler
Cylinder
Filter
Gasket
Guide vane actuator
Head
Packing case
Pipe
Pulsation vessel
Scroll
Seal panel
Separator
Shaft seal
Strainer
Unloader
Valve cover
Volute

Shutdown method
Interlock initiated
Manual shutdown

Split Casing Type
Horizontal
Vertical

Pressure test media/ type
Helium test
Hydro test

Impeller Type
Closed
Open
Semi-open

Manufacturer- In-line Centrifugal Compressors
Cooper
Delaval
Demag
Dresser Clark
Ingersoll Rand
Nuovo Pignone
Sulzer
Worthington

Anti-Surge method
Automatic
Manual
None

Balance Piston Fixation
Shrink-fit
Other

Fabrication Type
Cast
Milled
Welded

Labyrinth Type
Fixed
Rotating

Bearing Housing Type
Integral
Separate

Bearing Manufacturer
Glacier
Kingsbury
Other
Waukesha

Tilt-pad Bearing Load Orientation
Between pads
On pad

Tilt-pad Bearing Pivot Location
Center
Other

Thrust Bearing Lubrication
Directed
Flooded

Thrust Collar Type
Integral
Shrunk-on

Industry Standard
API617
API618
API619
API672
None
Other (specify)

Driver Type
Electric Motor
Engine
Expander
Gas Turbine
Other (specify)
Steam Turbine

Transmission Type
Direct connected (compressor same speed as driver)
Gear
Belt Drive

Shaft Sealing
Carbon ring
Combination
Dry face gas seal
Labyrinth
Mechanical
Oil seal
Other

Radial Bearing Type
Cylindrical
Elliptical
Magnetic
Multi-grooved
Offset halves
Other
Rolling element
Stepped
Tapered land
Three lobe
Tilting pad

Thrust Bearing Type
Flat face
Magnetic
Other
Rolling element
Stepped
Tapered land
Tilting pad

II.6. Pick Lists Related to Heat Exchangers

Equipment Type
Double pipe
Shell and tube
Fin-fan
Plate
Air cooled
Radiator
Dephlegmator
Multi-tube
Scraped surface
Wiped film
Spiral
Ambient air vaporizer
Steam vaporizer
Reboiler
Condenser
Other

Applicable Code
API Std. 660
API Std. 661
API Std. 662
ASME Section I (Power boilers)
ASME Section VIII (Pressure vessels)
CGA
CGA S1.3
None
Other
TEMA

Failure Description
shell rupture
tube rupture
cracked shell
cracked tube
tube fouling
partial pluggage
total pluggage
undersized
oversized
equipment damage
structural deficiency
separation at gasket

APPENDIX

III Procedure for Developing System-Level Taxonomies

The CCPS Plant Equipment Reliability Data Project sponsor companies, through representatives on a steering committee, are actively participating in the design and implementation of an industry-wide database. Part of the database development effort requires that members from the sponsor companies participate in the creation of the system-level taxonomies. The process of developing taxonomy information and data fields can be complex. This procedure was formulated to guide this effort in a rigorous, systematic fashion in order to ensure consistency among the various groups working to develop data field and taxonomy specifications.

This paper presents the background material along with the procedure to develop the information necessary to develop the taxonomies for new systems being added to the CCPS database.

III.1. Background and Theory

The Equipment Reliability Database is intended to collect data for many diverse types of equipment. The diversity of equipment in the database demands that equipment-specific fields are created to collect the appropriate data. The development of these equipment-specific data fields for all possible equipment is an arduous task. In order to expedite this task, the steering committee is contributing resources to develop the taxonomies and data field specifications for several equipment types.

III.1.1. Objective

The primary objective of taxonomy development is to produce a set of tables and pick lists that provide information to database programmers on how to add a specific system to the database. This procedure is designed to guide the user to successfully complete a series of tables that are used to program the data collection software for any equipment system

III.1.2. Overview of the Taxonomy Development Process

The taxonomy development process consists of several steps that are very iterative in nature. The steps are shown in Figure III.1 as a series of tasks; however, in practice there is some amount of iteration to modify and refine tables.

All reliability parameters are, in some form or another, a measure of the proportion of failures occurring in a population of equipment items. The goal of any taxonomy development process, therefore, is to capture two critical pieces of information whenever an item fails:

1. information regarding the nature of the failure (the "numerator" data)
2. information regarding the population of items susceptible to the failure (the "denominator" data)

To successfully collect both pieces of data, the overall steps in the taxonomy development process are as follows:

1. Define the system
2. Perform a functional analysis to identify and define failure modes
3. Specify the inventory-related fields
4. Develop event data fields, creating and documenting the rule set needed to link event data fields/combinations to failure modes or successes

The ultimate goal of the taxonomy development process is the creation of five key documents:

1. A **boundary diagram** for the system and a series of tables describing the data fields needed to collect inventory data for characterizing the equipment system.
2. Text and tables describing the **functional analysis** carried out to determine the failure modes for the equipment system. This document should also include the rule set used to determine each failure mode.
3. A series of tables specifying the **link between event data and failure modes**.
4. **Pick lists** for the data fields specified for **inventory data** (for example, list of manufacturers).
5. **Pick lists** for the data fields specified for **event data** (for example, list of events applicable to relief valve maintenance).

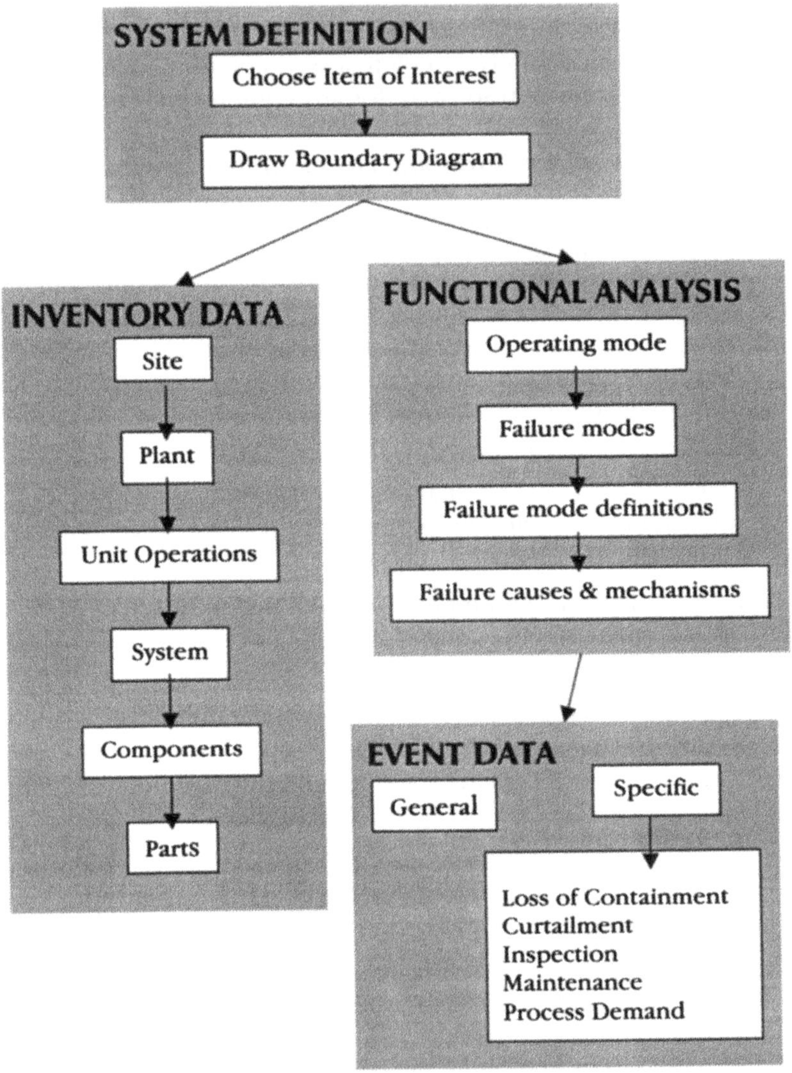

FIGURE III.1. Details of taxonomy development process.

The taxonomy development process is shown in detail in Figure III.1. The following sections provide instructions on how to create a taxonomy and data field specification for a new system. Examples using compressor systems and relief valves will be provided as a sample of how the taxonomy should be developed and presented.

III.2. Procedure

This procedure presents step-by-step instructions for finding the information needed to specify the complete taxonomy for a system. The overall steps in the taxonomy development process are as follows:

1. Define the system
2. Perform a functional analysis to identify and define failure modes
3. Specify the inventory-related fields
4. Develop event data fields

III.2.1. Define the System

The first step in the taxonomy development process is to define and categorize the system for which the taxonomy will be applied. Once the system has been defined, the boundary diagram can be drawn to illustrate the scope of the system itself. Part of the system definition includes the declaration of subordinate systems, as described in Section III.2.1.3.

III.2.1.1. Categorize the Item of Interest

In the CCPS data hierarchy, the system appears at the fourth level of indenture. The documentation for data hierarchy is contained within Chapter 5 of this Guidelines book. Before beginning work on the actual data field specifications, the taxonomy development group should review the current top-level taxonomy and prepare the recommended additions for the system of interest.

At the system level, there is a "subtaxonomy" that aids in specifying the system. While not part of the explicit taxonomy described in Chapter 5, it helps the analyst categorize the system of interest:

1. equipment grouping
2. equipment system
3. equipment type (if necessary)
4. equipment subtype

The equipment grouping refers to one of the classes of equipment in the CCPS taxonomy. Examples of groupings are heat exchange, relief devices, and rotating machinery.

The equipment system name is the specific name associated with the system of interest. In the case of the rotating machinery group, there are six different systems, such as pump, turbine and compressor, agitator, etc.

Some equipment systems can be further classified by type. An example of this is the rotating equipment system, which has types of compressors such as centrifugal or reciprocal. Further divisions among types may be needed, ***provided there is a separate industry specification to support the subtype.*** A good example of this occurs with centrifugal compressors, where in-line compressors have specification sheets different from integral compressors.

III.2.1.2. Draw the Boundary Diagram

The system boundary defines the subordinate systems and components that are to be included in the system. Keep in mind that, at some later date, separate failure modes and causes will be developed for each component appearing in a system. As a starting point, it is helpful to refer to boundary diagrams shown in the ISO proposed standard, OREDA, or a similar database. A few criteria are presented to assist in drawing a suitable boundary diagram:

1. For a component to be included, it must be necessary for the equipment to produce the intended function(s)
2. If a component is also another system, it is necessary to establish a parent/child relationship to allow proper analysis of either system.
3. It is not necessary for the component to be present in all applications in order for it to be included in the system boundary

The boundary diagram for the compressor is shown in Figure III.2. Note that the boundary diagram for the system satisfies Criterion 1; that is, all components shown within the boundary line are necessary for the compressor to deliver a certain quantity of fluid at a specified temperature and pressure, and without undue leakage to the atmosphere.

Note that interstage conditioning (cooling) is contained within the compressor boundary diagram and is also considered part of the equipment system for heat exchangers. In this case a parent child relationship exists, the parent being compressor and the child being heat exchanger.

Some of the equipment inside the boundary, for example the utility functions such as seal oil, may not be present for specific compressor applications, or will be present in vastly different levels of complexity. This does not merit their exclusion, as per Criterion 3.

III.2.1.3. Treatment of Subordinate Systems in the CCPS Database

Some systems, such as heat exchangers, motors, or control valves can function as an independent system in some applications; or as a support system in other applications. When supporting a larger system, the supporting systems function more like components of the larger system, rather than sys-

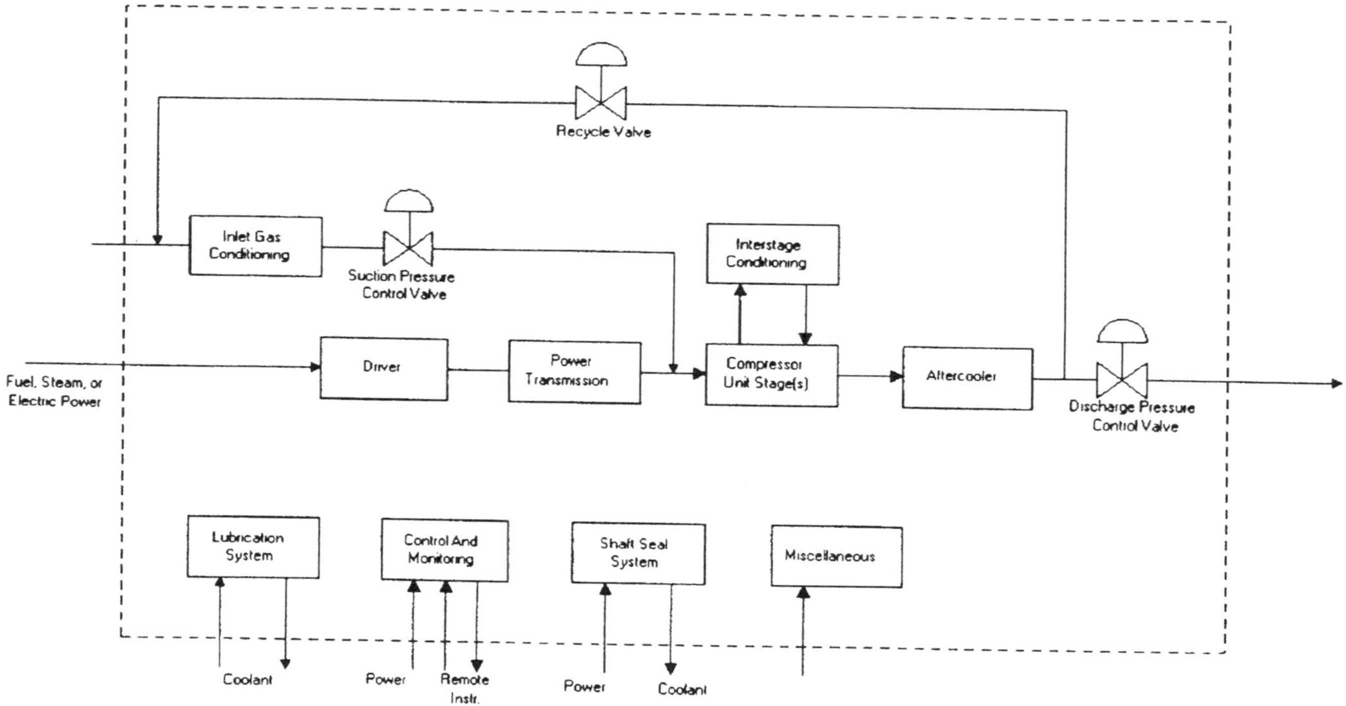

FIGURE III.2. ***Boundary diagram for compressor.***

tems themselves. These supporting systems are referred to as ***subordinate systems*** when they function as components for a larger system.

To resolve this dilemma, the Equipment Reliability Database has included the option to record the subordination of systems, so that failure of a subordinate system accurately cascades to failure of the primary system. The systems are registered as independent systems, but they are tagged as being subordinate to a larger system. Using the example of a heat exchanger supporting a compressor, the user would enter the database in the compressor system and would then be automatically linked to the heat exchanger system. The heat exchanger would appear as a component when system-level events are recorded.

The taxonomy development task does not require any input regarding subordinate systems. The dependencies between systems is recorded as systems are registered with the database.

III.2.2. Analyze Functional Characteristics

The analysis of functions and failures is a critical step in the taxonomy development process. The key result of this process is the listing of system failure modes. Failure modes form the foundation for collecting event data and estimating failure rates. Failure characteristics are derived from a four-step process (see Figure 2.2):

1. Identify functions and failures.
2. Determine the failure modes.
3. Specify failure modes in terms of event information.
4. Specify failure causes.

III.2.2.1. Identify the Functions and Failures

Before proceeding directly to failure modes, it is important that the analyst understands the nature of the functions and failures for the system. According to Rausand and Øien (Reference 3), "without evaluating failure causes and mechanisms it is not possible to improve the inherent reliability of the item itself."

This step is carried out in two smaller steps as follows:

- Review and document operating modes.
- Determine the functions.

III.2.2.2. Determine Failure Modes

As a first step in finding failure modes, all operating modes for a system should be considered. The term *operating states* refers to the state an item is in when it performs its intended functions. An item may have several operating states. Since each mode has different functions (and functions lead us to the definition of failures), all states should be reviewed when developing the data field specifications for an equipment item.

For instance, a valve may have several operating states. A process shutdown valve may have five operating states:

1. open
2. closing
3. closed
4. opening
5. modulating

States 1 and 3 are stable states, while states 2 and 4 are transitional. The essential function of this valve would be to close flow, that is, to operate successfully in states 2 and 3. It is necessary to consider all operating states to avoid overlooking some of the nonprimary functions of an equipment item.

A sample of operating states and modes for heat exchangers is shown below.

Operating Modes for Heat Exchangers

Running —Continuously exchange heat
Standby—Exchange heat on demand
Cyclical—Alternating service
Batch—exchange in batch mode

Operating States for Heat Exchangers

Exchanging heat
Not Exchanging Heat

After finding the operating states, you must determine the various functions of the system. The following scheme is recommended for development of the database field specifications:

- ***Primary***—the functions required to fulfill the intended purpose of the item. Simply stated, they are the reasons the item appears in the process.
- ***Auxiliary***—functions that support the primary function. Containing fluid is a typical example of an auxiliary function for a compressor. In many cases, the auxiliary functions are *more critical* to safety than the primary functions.

- ***Protective***—functions intended to protect people, equipment, or the environment
- ***Information***—functions comprising condition monitoring, alarms, etc.
- ***Interface***—functions that apply to the interface of the item in question with another item. The interface may be active or passive. For example, a passive interface is present if an item is used as the support base for another item.

As noted in the referenced article, ***primary functions*** are "the reasons for installing the item." Hence, the primary functions listed above pertain to changing a fluid's pressure and flow rate for the compressor example. Auxiliary functions, such as containing fluid, are classified as such since these functions support the essential function of increasing flow rate or relieving pressure. They are not deemed to be less safety critical, however.

The process is repeated for all components and then all parts associated with the various components. An example is provided below from the heat exchanger taxonomy documentation.

System Type: [Heat Exchanger without Change of Phase]

Class of Function	Function Description
Primary	Provide correct heat exchange at a desired rate
Auxiliary	1 .Contain cooling and heating fluids 2. Prevent mixing of cooling and heating fluids 3. Allow correct fluid flow rates and distribution
Protective	• Prevent damage of downstream equipment • Prevent damage to heat exchanger equipment
Information	Condition monitoring
Interface	None

The next step is to find the failure modes based on the functions declared in the previous step. The functions defined above now become the basis for the heat exchanger failure modes.

As an aid in finding failure modes, it is useful to think of classes of failure modes. These classes capture the information about the ***timing*** and ***severity*** of the failure mode. The following classes of failure modes have been defined:

- ***Intermittent***—failures that result in a lack of some function for a very short period of time. The item will revert to its full operational standard immediately after the failure
- ***Complete and sudden***—failures that cause a complete loss of a required function and cannot be forecast by testing or examination. These are typically referred to as *catastrophic* failures.
- ***Partial and sudden***—failure that does not result in a complete loss of function and cannot be forecast by testing or examination.
- ***Partial and gradual***—failure that does not result in a complete loss of function and could be forecast by testing or examination. These are commonly known as *degraded operation.* A gradual failure represents a "drifting out" of the specified range of performance values. The recognition of gradual failures requires comparison of actual device performance with a performance specification.

To illustrate the nature of the failure mode information required, a portion of the failure mode table for heat exchanger is given below.

Class	Function	Operational Mode	Failure Mode	Failure Category	Failure Mode Definition
Primary	Provide correct heat exchange capacity	All modes	Total loss of cooling	Complete and Sudden	Total absence of heat transfer
			Reduced cooling	Partial and Sudden	Duty reduced to below that needed to maintain intended operation
			Reduced cooling	Partial and Gradual	Duty reduced to below that needed to maintain intended operation

III.2.2.3. Specify Causes

A failure cause is defined in IEC 50(191) as " the circumstances during design, manufacture or use which have led to a failure." For the sake of data field specification, review and record failure causes for each cause category. Use causes from the pre-defined list wherever possible. If the list does not contain a necessary cause, then record the new cause. Upon sponsor approval, the new causes will be added to the CCPS database for programming.

III.2.3 Specify the Inventory Data

Inventory information consists of data needed to classify or categorize equipment for specific filters or searches. It also is needed to provide the population basis for the estimation of equipment failure rates.

The seven levels for the database taxonomy are as follows:

1. Industry
2. Site
3. Plant
4. Unit operations
5. System
6. Component
7. Part

The taxonomy development process will determine the make up of the taxonomy for Levels 5–7. It is important, therefore to define the scope of the system to be addressed as it fits into taxonomy levels 1 - 4 as part of the first step in this procedure.

III.2.3.1. Revise the Tables and Pick Lists for General Inventory-Related Fields

A number of fields have set pick lists that may need revisions or new entries, based on the findings of the analyst. Check the tables in the list below to see if any changes need to be made. Pick lists are specified in Appendix II.

1. Review the current General System Inventory Table (applicable to all types of equipment) to determine if the new system type requires any new field or entries for Pick Lists
2. Update the Equipment Class Table (SYS2 and SYS2.1) to include the new equipment type
3. Review the Operating Mode (SYS3) tables to determine if modifications are necessary to accommodate the new equipment type
4. Review and update (if necessary) the table for variable SYS4 – External Environment
5. Review the table for Environment—Vibration (SYS6) and make changes
6. Review and Update the table for SYS7—Intended Maintenance Strategy

III.2.3.2. Define Data Fields for System-Specific Inventory

The purpose of the system-specific inventory fields is to be sure that all characteristics that could influence the reliability of a given type of equipment are recorded. When proposing a new inventory field, a key question to keep in mind is "will this field provide information that could realistically distinguish between failure rates for this equipment?"

For instance, if failure rates for compressors were known to have no dependence on the power consumption of the compressor driver, it would be a waste of time to collect that information.

In order to simplify the specification of inventory fields, it is strongly recommended that the analyst first consider industry-standards for specifying the equipment in question. For instance, heat exchangers could use the fields specified in the heat exchanger specification sheet in standards of the Tubular Exchange Manufacturer Association.

The CCPS Equipment Reliability Database uses a set of predefined tables for inventory-related information that is common to all types of systems. These are referred to generically as the "EQP" tables, since their code name begins with the three-letter designator EQP. For instance, EQP tables have been established for the following system parameters:

EQP1A/B—Materials of Construction
EQP2—Connection Type
EQP3—Fluid Phase (liquid, gas, two-phase, etc)
EQP4—Corrosiveness
EQP5—Fluid Fouling Factor

If the taxonomy developer encounters the need to specify one of the above parameters, the developer must defer to one of the existing EQP tables, instead of creating a new table.

After you have created the necessary data fields, create the Pick Lists for any fields that require standard inputs. If, for example, the developer chose to include an inventory field entitled "Shaft Seal Type" the Pick List of possible seal types needs to be developed.

III.2.3.3. Define the Pick Lists for Inventory Fields

If needed, the pick lists for inventory fields should be defined and provided to the database along with the field specification. Be sure to create a variable name in the field specification, so that it can be linked to the correct pick list. *Do not create pick lists that have already been created. Refer to Appendix II for a listing of standard pick lists.*

III.2.4. Specify the Event Data

The event tables provide the link between the data collected in the field and the analysis of failures made in the software. Event data provides the information regarding the magnitude of the numerator in the calculation of reliability characteristics. This means that the event data needs to be collected to answer these and other questions:

- How was the failure detected?
- What happened to the equipment (failure mode)?
- How long was the equipment out of service?
- What caused the failure?

III.2.4.1. Specify the Applicable Events

The first piece of information required for the programmer is the listing of the types of failure events that are applicable to the equipment item in question. The analyst provides this information to the programmer by completing the rightmost column in following table, using a yes or no response. An example of the completed table for heat exchangers is shown below.

Pick List PLF1 **Type of Event**	**Subtype of Event**	**Applicable to Heat Exchangers?**
Loss of containment		Yes
Startup		No
Shutdown		No
Curtailment		Yes
Inspection		Yes
Maintenance	Proof test	No
	Repair following failure	Yes
	Preventive maintenance	Yes
	Condition-based maintenance	Yes
Process demand		Yes*

*In standby operating mode only

The answer "yes" in the preceding table tells the programmer to create a table of fields for that particular type of failure event. As you would expect, if a type of event is given a "yes" a corresponding table should be provided by the analyst.

All system events have five fields of data that are common to all types of events. These fields are assumed to exist in all event tables, so it is not necessary for the analyst to provide this in the event tables:

- Subscriber ID
- Site ID
- Plant ID
- Process Unit ID (optional)
- System ID

The following sections provide some pointers on developing field tables for each of the generic types of failure events.

III.2.4.2. Specify How to Determine Failure Modes from Event Data

One of the strengths of the CCPS database is that it has been designed to be as objective as possible. This has been achieved, to a large extent, by promoting the concept of inferring failure modes from raw event data, as opposed to relying on the opinion of a data collector.

While not an absolute requirement, it behooves the analyst to develop the structure to collect objective data from the field and determine failure modes by a fixed algorithm, embedded in the software. In order to program the software to make the determination of failure modes, the analyst must also supply a description of the algorithm needed to determine the failure mode, in terms of a flowchart (preferred by the programmers) or text.

To provide the linkage between failure modes and events, it is helpful if the analyst first prepares a table showing the relationships between failure modes and events. An example of this table is provided below for heat exchangers:

System Failure Modes	Failure Mode Definition	Relief Valve Failure Logic	
		Input Form	**Criteria**
Fail to open	Fail to open prior to reaching 1.5 times set pressure	Demand	Relief lift = 'NO' or lift ratio equal to or greater than 1.5
		Proof Test	Fail to open if "plugged" or lift ratio equal to or greater than 1.5 or max pressure equal to or greater than 1.5
		Inspection	Discharge piping plugged field = Plugged
Opens above set pressure	Opens between 1.1 and 1.5 times set pressure	Demand	Lift ratio between 1.1 and 1.5
		Proof Test	Lift ratio between 1.1 and 1.5

III.2.4.3. Specify the Fields for System-Specific Events

For each of the events below, it is necessary for the developer to review all fields and create or modify fields to be compatible with the system of interest.

System Event—Loss of Containment

Loss of containment events can occur for nearly all systems. A sample of the fields used for system-level loss of containment events is shown below for relief valves:

- Event date and time
- Event end date and time or
- Duration
- Pressure at time of failure
- Fluid Released
- Severity
- Quantity Released, lbs
- Hole Size
- Estimated Area
- Event description
- Consequence
- Reportable Incident?
- If reportable, reportable to:
- Comments

System Evet—Shutdown

If a system is shutdown, certain data are recorded for the estimation of reliability parameters. The shutdown event ids typically used to record a shutdown of a system that is not directly attributable to a failure mode of that system. For instance, a shutdown may occur due to a power failure, but this would not be recorded as a failure of the compressor.

- Event date
- Event time
- Date and time returned to service
- Is there a force majeure? Y/N
- Shutdown cause
- Shutdown subcause
- Scheduled activity? Y/N/Unknown

System Event—Startup

The system startup event is characterized by ability of the system to start according to predefined criteria. A failure to start is, by itself, a failure mode for most systems. If the field for successful start is a no (N), the failure mode is automatically set to "fails to start."

System Event—Curtailment

Curtailments events are usually linked to the "partial loss of throughput" failure mode. The following is a sample of fields appearing in a System Curtailment table.

- Event start date and time
- Event duration, or
- Date and time returned to full capacity
- Cause of curtailment
- Maintenance Required?
- Average percent capacity available

System Event—Inspection

Inspections typically do not mean that a failure has occurred; however, they can point to a failure mode that may not be easily detected. The Inspection Event table records the necessary information from a system inspection. For relief valves, a sample of the data fields are provided below.

- Discharge piping visual inspection
- Weep hole visual inspection
- Evidence of seat leakage
- External leakage
- Repair required
- Pressure between rupture disk and relief valve
- Insulation loss?
- Heat tracing operating?
- Pressure monitor functional?
- Temperature monitor functional?
- Excess flow check valve functional?

System Event—Process Demand

Process upsets may trigger failures or successes for some systems. A partial listing of the process upset data collected for relief devices is provided in the table below.

- Event end date
- Event duration or event end time (Depending what is picked, the other should be calculated)
- Reason for process demand
- Event description
- Valve lift? (Discharge through the relief valve)
- Lift Pressure
- Maximum process pressure achieved
- Rapid cycling of the valve (chatter) audible during incident?
- Valve reseat?
- Reseat pressure
- Code deviation?
- Protected equipment damaged?
- Loss of containment?
- Repair or replacement required?
- Proof test performed
- Additional information

System Event—Repair

Repair events occur when a repair is made following a failure, proof test, inspection, or condition-based maintenance. The following fields are part of the System Repair Event table for relief devices.

- Maintenance Start Date
- Maintenance Start Time
- Date maintenance completed
- Maintenance hours expended
- Maintenance performed
- Components repaired
- Components replaced
- Part replaced
- Component failure mode
- Component failure cause(s)
- Part failure mode
- Part failure cause(s)
- Additional information
- Disposition following repair

System Event—Proof Test

The fields used to record data for a relief device Proof Test Event are shown below.

- Proof Test Date
- Event Time
- Visual Flow Path Inspection: Inlet
- Visual Flow Path Inspection: Outlet
- External leakage
- Seat Leakage
- Leak Pressure
- Valve lift?
- Lift Pressure
- Maximum test pressure
- Valve reseat
- Reseat Pressure
- Leakage following reseat
- Maintenance Performed?
- Actual test time (hours)
- Additional information

System Event—Calibration

The fields used to record data for a calibration are shown below.

- Calibration date
- Time of calibration
- Calibration performed in-situ (Y/N)
- As-found value(s)
- Point check (P) or range (R)?

References

1. IEC 50(191), International Electrotechnical Vocabulary (IEV), Chapter 191: Dependability and Quality of Service. International Electrotechnical Commission, Geneva, 1990.
2. OREDA-1992, "Offshore Reliability Data." DNV Technica, Høvik, Norway, 1992.
3. Rausand M. & Øien, K., The basic concepts of failure analysis. ***Reliability Engineering and System Safety*** **53, pp. 73–83** (1996).

Index

N

O

P

Q

R

S

Publications Available from the
CENTER FOR CHEMICAL PROCESS SAFETY
of the
AMERICAN INSTITUTE OF CHEMICAL ENGINEERS

CCPS Guidelines Series (continued from page ii)

Guidelines for Technical Management of Chemical Process Safety
Guidelines for Chemical Process Quantitative Risk Analysis
Guidelines for Process Equipment Reliability Data with Data Tables
Guidelines for Safe Storage and Handling of High Toxic Hazard Materials
Guidelines for Vapor Release Mitigation

CCPS Concepts Series

Inherently Safer Chemical Processes. A Life-Cycle Approach
Contractor and Client Relations to Assure Process Safety
Understanding Atmospheric Dispersion of Accidental Releases
Expert Systems in Process Safety
Concentration Fluctuations and Averaging Time in Vapor Clouds

Proceedings and Other Publications

Proceedings of the International Conference and Workshop on Reliability and Risk Management, 1998
Proceedings of the International Conference and Workshop on Risk Analysis in Process Safety, 1997
Proceedings of the International Conference and Workshop on Process Safety Management and Inherently Safer Processes, 1996
Proceedings of the International Conference and Workshop on Modeling and Mitigating the Consequences of Accidental Releases of Hazardous Materials, 1995.
Proceedings of the International Symposium and Workshop on Safe Chemical Process Automation, 1994
Proceedings of the International Process Safety Management Conference and Workshop, 1993
Proceedings of the International Conference on Hazard Identification and Risk Analysis, Human Factors, and Human Reliability in Process Safety, 1992
Proceedings of the International Conference and Workshop on Modeling and Mitigating the Consequences of Accidental Releases of Hazardous Materials, 1991

Printed and bound by CPI Group (UK) Ltd, Croydon, CR0 4YY
16/09/2026
14963420-0002